"We are like a little child entering a huge library. The walls are covered to the ceilings with books in many different tongues. The child knows that someone must have written these books. It does not know who or how. It does not understand the languages in which they are written. But the child notes a definite plan in the arrangement of the books - a mysterious order which it does not comprehend, but only dimly suspects"

Albert Einstein

WHY I BELIEVE GOD EXISTS
Discourse on the scientific evidence

Andrew C. Phiri

Copyright © 2015 Andrew C. Phiri
www.andrewcphiri.com

PO Box 37919
Lusaka, Zambia.
voiceoftheword@live.com

Printed in the USA for worldwide distribution

All rights reserved. No part of this publication may be reproduced or transmitted in any form or by any means, electronic or mechanical, including photocopying, recording, or by any information storage and retrieval system, without permission in writing from the publisher. To duplicate this book without permission is a violation of international copyright laws.

ISBN-13: 978-1503121300
ISBN-10: 1503121305

Acknowledgements
Unless otherwise indicated all scripture quotations were taken from the Holy Bible - King James Version.

NOTE on citations: A name followed by date in parenthesis indicates the source of the referenced text. Find a matching entry in *References* for full details of the item. Text in square brackets embedded in quoted words are author's comments.

This book is available for purchase on Amazon, eBay and many other e-stores.

To see news updates visit and like the Facebook page:
www.facebook.com/andrewcholaphiri

Cover design: Hubert D'souza

To
Norah, my love, Mishael and Terry, beloved sons;
a family that is of immense joy and encouragement.

CONTENTS

Note		12
Introduction		14
1.	RELIGION AND VIOLENCE *What really lies beneath?*	19
2.	PROOF OR EVIDENCE? *Getting our terms right*	35
3.	ORIGIN OF THE UNIVERSE *Which way does the evidence lead - God or Chance?*	44
4.	FINE-TUNED UNIVERSE *How come life?*	55
5.	THE INFORMATION EVIDENCE *Whence came information?*	67
6.	THE MORAL ARGUMENT *How come the moral compass?*	79
7.	PAIN AND SUFFERING *Just where is the benevolent creator?*	90
8.	IN SEARCH OF GOD *Have you the right attitude?*	103
Questions and Answers		117
Recommended readings		146
Appendices		147
References		156
Index		160

Questions and Answers

Questions atheists and agnostics have asked through email correspondence and posts on my blog. Find my responses on the page numbers indicated below.

Question	Page
1. *"If you were born in Afghanistan, you would be a Taliban and would have bombed for your God."*	117
2. *"We still do not know how life started but we know how it progresses. We are working on this one, but we won't substitute our not knowing for something delusional."*	119
3. *"Adam and Eve were on earth 6000 years ago according to chronological data presented in the Bible, but archaeology and palaeontology have revealed fossils of dinosaurs and apes which date back to millions of years ago. So, hasn't science made the Bible obsolete?"*	121
4. *"Give me the evidence that a designer exists and I will solemnly change my views."*	127
5. *"Tell me, what benefit is in forcing people (taking children to church at an age when they aren't ready to decide) to attend church service which will force them to worship a god they have*	129

not seen?"

6. *"The Bible is a white-man's book. Do you know what most whites have done after bringing religion to Africa? They have abandoned it, because it has done its job of being a mass enslavement tool."*   131

7. *"If God exists why are there different religions; Christianity alone has thousands of denominations?"*   132

8. *"Religious institutions do not pay tax, yet they are able to engage in political activities, public policies and everything to do with government. The money that religion alone has collected has the power to help a lot of people."*   134

9. *"Religion has been practiced for thousands of years and clearly it has not yielded any good results, as it has failed at achieving peace except of course robbing people of their little money by selling them delusions of heaven and hell…"*   135

10. *"…even if you say that religion has brought in good things. What things? Money?? Money is not what people need, people need mental freedom to exist."*   138

11. *"People become Christians because they fear hell fire and want to escape it by going to heaven. All these are delusions."*   139

12. *"One thing you must also know is that, it is not possible for a supernatural to come in to a natural world."*   140

13. *"Why do we need God when evolution has explained how things became what they are?"*   140

NOTE

On the terms "God" and "Religion" as used in this book

It has been scientifically observed that the universe is stretching with galaxies flying away from each other. Thus, rewinding time would takes us to a place where all matter and energy are heavily condensed in a single spot called the *singularity*. Rewinding further would thin out all matter, space and time into *nothingness*. This therefore implies that the transcendent power that caused the first particle to burst forth into existence had to be itself non-physical and timeless. However, this power could not only be a force or it would never have had the capacity of bringing forth creation. Moreover, one of its products were human beings endowed with sophisticated intelligence; creatures which have explored the physics of the universe and have been bewildered at the fine-tuning of its fundamental constants. It therefore goes without saying that the creator is a far more superior intellect who spoke - "*Let there be…*" - and it *was*. This timeless, non-physical, all-powerful and intelligent being is what is referred to as God in this writing.

Some people do not believe in the existence of God; they are called *atheists* or simply *unbelievers*. Others do and are called *theists* or simply *believers*. Believers in turn have various conceptions of God and this has given rise to different religions as well as many other factions within particular religions. In this book, the word religion is not used in that sense; it is used in a general context of meaning 'belief in God'. My aim is to give general arguments for the existence of God without getting into particular doctrinal issues of explaining which religion, or holy book, or denomination is the true one. However, I speak as a Christian, not merely for the sake of it, but because I find the Christian thought about God to be more coherent and rational.

INTRODUCTION

In February, 2015, a Zambian atheist-medical practitioner and author of *The Sceptic* - Dr. Teddy Mulenga - stirred a hornets' nest when he featured on a TV programme during which he presented some of his atheistic views. This was too controversial for a deeply religious Zambian audience, unaccustomed to courage that questions the existence of God. However, unknown to many, the God-debate has actually been silently warming up in the country (especially in universities among students of philosophy and also on social media) and Dr. Mulenga was only a brave one to stir up the controversy. There could be many people who share Mulenga's skepticism but prefer to remain mute either for fear of religious intolerance or simply finding it a futile venture. Much needed now are Christians who are well versed in Bible truth and who can also engage in intellectual discourse to answer unbelievers' questions even as we have been admonished in the Scripture to *"be ready always to give an answer to every man that asketh you a reason of the hope that is in you with meekness and fear"* (1 Peter 3:15). This verse informs us that the Christian hope should be grounded in reason and not blind faith!

When Mulenga expressed his views about Faith, I expected spiritual people to respond with matching intellectual bravery, but sadly much of what came up was mere rhetoric. What is more, some clerics even went to extents of threatening the atheist with *hell-fire*. But just how effective is the threat of hell on a person who doesn't even believe it exists? I am afraid that this situation makes even the most naïve atheist feel smart (and even 'holier' in cases where some religious people have resorted to abusive language in expressing their disgust at atheism).

Now, although the God-debate may appear strange in Zambia, a very religious nation which cherishes its Christian declaration, it is important for Zambian believers to be aware that the debate has been very active and rigorous in Europe, Australia and the USA. It is important to be aware that atheism and secularism is on the rise. *YouTube* and a plethora of blog sites are replete with various engagements between prominent atheists (like Richard Dawkins, Sam Harris, Daniel Dennet and the late Christopher Hitchens) and renowned *theists* (like William Lane Craig, John Lennox, Alister McGrath, Francis Collins, and John Polkinghorne). If Christians in places where the God-debate is unknown choose to close their ears to issues raised by atheists, they sure will be disappointed to later wake up to the reality of a new crop of youngsters who have already been exposed to atheism through university education or the internet. An ostrich approach of 'burying' the head to avoid a problem has never worked at any time. A person with truth does not fear to face and verify falsehood for *"God hath not given us the spirit of fear; but of power, and of love, and of a sound mind"* (2 Tim.1:7). Of big concern is, among many modern believers are there those with *"a sound mind"*, equipped with the ability to face the opposition as they earnestly give heed to saint Paul's admonition to *"Study to show thyself approved unto God, a workman that needeth not to be ashamed, rightly dividing the word of truth"* (2 Tim.2:15)?

On my bookshelf are writings by both atheist and theist professors. It has been quite informative to read and cross-examine the diametrically opposed worldviews. This far I have found the theistic worldview to be more logically coherent and convincing. And in reading and hearing what many renowned atheists have had to say, I cannot agree more with the theologian Paul Copan who noted that most atheists do have expertise in certain fields like biology or evolutionary theory, but they turn out to be fairly disappointing when arguing against God's existence (Copan, 2011).

In a nutshell, the following are usually the main arguments atheists often present against the case of the existence of God:

- *"Faith is evil as it has brought more harm to society - there would never have been 9/11, ISIS or Boko Haram had religion not existed"*.

- *"Postulating God as the creator only raises the question of*

'Who created the creator?'"

- *"There is no scientific evidence for the existence of God".*

- *"How can a perfect God create an imperfect world filled with the evil of pain and suffering?"*

In this book I present a cumulative-case discussion which shows that there are good reasons for believing in the existence of God. I begin with the first item in the above list – the discrediting of religion on the basis of the violence committed in its name. Many people are increasingly getting convinced that faith is a vice and not a virtue. A closer look at this problem will show that while this may appear to be so, there is actually a different underlying factor that incite suicide bombings and other acts of terrorism around the world.

1.
Religion and Violence
What really lies beneath?

"His eyes met mine. They were deep brown, nearly black. They struck me as wise and sad. He turned back toward the New York Stock Exchange building. With a long gnarled finger, he pointed at the building.
'Or maybe', I agreed, 'It's about economics, not religion'"

John Perkins

"Without religion, we would still have the twin towers in USA standing. There would never have been the fanatical Osama Bin Laden. The many civilians dying in the towns of Nigeria, at the hands of Boko Haram would be alive today...What about the 200 children abducted in Nigeria and held captive at the hands of a religious fundamentalist group Boko Haram? Which religion would allow such cruelty? Surely God must be having his diary full! He surely deserves a vacation!" - Words of Dr Teddy Mulenga, a Zambian atheist medical practitioner, in his book *The Sceptic*.

Seemingly a good case against religion, but not factually true. If one is to be as naïve about the real causes of violence in human societies, he or she may as well say the following concerning science:

- *Without the Scientific equation of $E=MC^2$ there would never have been an invention of an atomic bomb which destroyed Hiroshima and Nagasaki.*

- *Without the technology of science there would never have been the Air Asia aircraft which plunged into Java sea killing people.*

- *Without Internet technology we would never have had cybercrime which is responsible for huge losses of millions of dollars every year.*

An informed person should know that science in itself, or, faith in God, has never been harmful; on the contrary it is the abuse of science and religion that is evil. And while science can be abused, we should never lose sight of the greater good that has come out of it. Likewise, whenever there is a natural (or some other) disaster, is it not the churches that are usually first to organize relief food and shelter? Count orphanages and note how many are owned by atheists.

Now, a few words about the alleged relationship between faith and violence: Is there truly a *causal relation* between faith and the on-going terrorist suicide bombings and insurgents in the Middle East? "Yes" is the simple answer to give when the situation is casually examined, but research facts show such an answer to be simplistic and myopic. Here is an illustration to help explain this: if a scientific method is used to assess facts, two events stand in a causal relation when event B is caused by A. For this relationship to be valid there should be consistent evidence that A causes B. For example, in 1998 Alexander Fleming, a Scottish scientist, serendipitously noticed that a mould which had grown on one neglected dish was killing bacteria. Repeated experiments demonstrated that the presence of this mould (called *Penicillium notatum*) caused bacteria to die. This causal relation later led to the creation of the first antibiotic drug, penicillin. This method of establishing facts by observing causal relations is also applicable in other research disciplines such as economics. For example, demographic data shows that better living standards always lead to an increase in demand for resources – let a community move into the middle or higher income range and the next thing will be more cars on the roads; this will directly translate to more demand for fuel. This causal relation is so valid that economists can reliably predict future levels of demand for resources. But now, can we confidently bring out facts which show a causal relation between faith and violence? A simple analysis is all that is needed: First let us identify things that will work as variables for increasing or decreasing *faith*. Common in most religions would be such things as *prayers, fasting, reading scripture,* and *gathering for fellowship,* just to mention four. Now, if a causal relationship exists between faith and violence, it should be expected that the fluctuations of these variables should correspond to (or at least show some signs of effect on) the levels of violence committed in the name of faith. Let us look at one case, a well-known situation that led to the formation and rise of al Qaeda. The questions to ask are: what really inspired the rise of this organisation and what agitated its various resulting insurgents? Was the formation of al Qaeda preceded by intensive prayers and reading of the Quran? Certainly not; preceding the formation of al Qaeda was the Soviet Union's invasion of Afghanistan!

The invasion of Afghanistan and the rise of al Qaeda

Afghanistan used to be led by a communist soviet-supported government. This government was always in antagonism with anti-communist fighters. In April 1978 Afghanistan's government was overthrown by left-wing military officers. The new government which was quite an unpopular one forged ties with the Soviet Union as it embarked on a ruthless campaign to supress all domestic opposition. The new ambitious government started vigorous land and social reforms that were vehemently opposed by the largely anti-communist and Muslim population. This provoked insurgencies and led to the stormy rise of what became known as the *mujahideen*, an Arabic word for "*those who engage in jihad*". The chaotic situation stirred the Soviets to invade Afghanistan on 24th December 1979. With the United States' backing of the mujahideen, the rebellion spread country-wide. Notes Britannica Encyclopaedia (2015):

> *For the mujahideen the quality of their arms and combat organization gradually improved, however, owing to experience and to the large quantity of arms and other war materiel shipped to the rebels, via Pakistan, by the United States and other countries and by sympathetic Muslims from throughout the world*" and "*In addition, an indeterminate number of Muslim volunteers—popularly termed 'Afghan-Arabs,' regardless of their ethnicity—travelled from all parts of the world to join the opposition.*

This became a long and costly war which later saw the Soviet Union withdraw its forces sometime in 1988. While there was a palpable sense of relief and victory among the people and the thousands of volunteer militants, for the leaders of mujahideen another phase of a world-wide war had just began. Between August and September of that year, strategic meetings were held that would lead to the military training of many foreign volunteers. These would be trained in insurgent and terrorist tactics whilst in Afghanistan and be later sent into the world to pursue their agendas (Stern and Berger, 2015). Out of this large pool of fighters, the best and elite were to be admitted into an inner exclusive circle that was to form 'the base' of the new revolution. 'The base' in Arabic language is the word 'al Qaeda'.

Another case of Iraq

Following the United States invasion of Iraq, terrorism rose precipitously in the country (Stern and Berger, 2015). Nada Bakos, the CIA officer who was charged with taking down the leader of the terrorist organisation in the country - Zarqawi – noted the following:

> *The war in Iraq provided al Quaeda a new front for its struggle with the West… the United States didn't 'face-down' al Qaeda in Iraq; it inadvertently helped Zarqawi evolve from a lone extremist with a loose network to a charismatic leader of al Qaeda. By extension, it would be safe to say that the al Qaeda in Iraq affiliate, Jabhat al-Nusra, exists because of the Iraq invasion, and likely would find new authority and power if the United States made Syria the next front for the global jihadist movement*' (Bakos, 2013).

Many words can go on being written about various incidents that fuelled insurgents and other terrorist activities. In all the incidents the pattern that emerges is not that of a causal relation between 'more prayers' and 'increased insurgents' but 'political invasions/interventions' and 'insurgents'. This observation is supported by research works done by experts. Pape (2005) has shown that the fundamental motivation for people to get enthused into fighting 'a just cause' and killing anyone who smells 'infidel' is often political; the desire to force the withdrawal of foreign forces occupying land believed to belong to an oppressed people. An incident narrated by Perkins, author of bestseller *Confessions of an Economic Hit Man,* puts it well; it occurred to him some two months after 9/11:

> *The sights and smells were overwhelming – the incredible destruction; the twisted and melted skeletons of those once-great buildings...a man rushed out of an office... He had a scrawny grey beard and wore a grimy overcoat...I knew he was Afghan. He glanced at me...From the way he looked straight ahead, I realized it would be up to me to begin the conversation. "Nice afternoon."*
>
> *"Beautiful." His accent was thick. "Times like these, we want sunshine."*
>
> *"You mean because of the World Trade Center?"*
>
> *He nodded.*
>
> *"You're from Afghanistan?"*
>
> *He stared at me. "Is it so obvious?"*
>
> *"I've travelled a lot. Recently, I visited the Himalayas, Kashmir".*
>
> *"Kashmir." He pulled at his beard. "Fighting."*
>
> *"Yes, India and Pakistan, Hindus and Muslims. Makes you wonder about religion, doesn't it?"*
>
> *His eyes met mine. They were deep brown, nearly black. They struck me as wise and sad. He turned back toward the New York Stock Exchange building. With a long gnarled finger, he pointed at the building. "Or maybe", I agreed, "It's about economics, not religion"* (Perkins, 2006, p.190-192).

Precisely so! The McGraths are right:

> *The history of the twentieth century has given us a frightening awareness of how political extremism can equally cause violence. In Latin America, millions of people seem to have 'disappeared'*

> *as a result of ruthless campaigns of violence by right-wing politicians and their militias. In Cambodia, Pol Pot eliminated his millions in the name of socialism...the reality of the situation is that human beings are capable of both violence and moral excellence — and that both these may be provoked by world views, whether religious or otherwise...Suppose Dawkins' dream were to come true, and religion were to disappear: would that end the divisions within humanity? Certainly not. Such divisions are ultimately social constructs, which reflect the fundamental sociological need for communities to self- define, and identify those who are 'in' and those who are 'out'; those who are 'friends' and those who are 'foes'"* (McGrath and McGrath, 2007, p.49-51).

In analysing and commenting about what enthuses people from different countries to willingly volunteer in joining and fighting for a terrorist organisation, Stern and Berger (2015) make an important note:

> *For many, perhaps most, jihadists, religious motivations are necessary but not sufficient to explain the leap to violent action. Some mix of political sentiment, religious belief, and personal circumstance is required. Parsimonious explanations, which focus only on single external factors, whether religious or political, cannot explain why one sibling becomes a jihadist and another a doctor. Clearly, something happens that makes an individual willing to risk his or her life for a cause* (p.83).

But, what is this "*something*"? Eckert Tolle explains it well:

> *Fear, greed, and the desire for power are the psychological motivating forces not only behind warfare and violence between nations, tribes, religion, and ideologies, but also the cause of*

incessant conflict in personal relationships. They bring about a distortion in your perception of other people and yourself. Through them, you misinterpret every situation, leading to misguided action designed to rid you of fear and satisfy your need for more, a bottomless hole that can never be filled.

And, in diagnosing the condition of human history, Tolle presents a correct prognosis:

Chronic paranoid delusions, a pathological propensity to commit murder and acts of extreme violence and cruelty against perceived 'enemies' – his own unconsciousness projected outward. Criminally insane, with a few brief lucid intervals. (Tolle, 2005, p.11-12).

At the bottom of all this is the fact that human beings are creatures of ego; this ego can feed on religious or atheistic forms of ideology. And acts that will be committed in the name of that ideology will be a result of a nexus of factors including the inherent nature of self-defence in people, and the prevalent political and social influences.

To use an analogy, a person is like a container and a worldview like a liquid, say water. If the container is rusty and dirty, the liquid will take on the dirt and its colour. However it will be an error to define the liquid by the colour of the dirt. This can be said to be similar to the way white slave traders came to Africa as 'containers' of the Gospel but which was polluted with the greed of economic exploitation at the expense of human freedom. The Christian slave traders justified their immoral actions by misplacing and misinterpreting Old Testament Jewish practices of slavery. Note that just as various Biblical themes can be misplaced and misapplied in order to justify a goal, so can atheism or other worldviews. To emphasise McGrath's note on this again, *"the reality of the situation is that human beings are capable of both violence and moral excellence – and that both these may be provoked by world views, whether religious or otherwise".* Evil is inherent in the human nature and this nature can wittingly or unwittingly adapt and manipulate a worldview for its ends.

Old Testament violence?

It has often been said that if Joshua, Gideon, Samson, David and other various warriors of the Old Testament were living in modern times, they would be facing charges of crimes against humanity in the Hague.

While Christians have even come up with different hymns which reminisce the victorious acts of these men – remember the Sunday School rhyme, *"Joshua won the battle of Jericho"*? – Atheists get appalled at how believers condone the Biblical acts of 'genocide'; Israeli wars against non-Yahweh worshippers – Amalekites, Hittites, Philistines, e.t.c - are today said to be no different from ISIS' or Boko Haram's atrocious acts against non-Muslims. Joshua's destruction of Jericho is indistinguished from Hitler's invasion of Poland. This is Richard Dawkins famous description of the God of the Old Testament:

> *The most unpleasant character in all fiction: jealous and proud of it; a petty, unjust, unforgiving control-freak; a vindictive, bloodthirsty ethnic cleanser; a misogynistic, homophobic, racist, infanticidal, genocidal, filicidal, pestilential, megalomaniacal, sadomasochistic, capriciously malevolent bully"* (Dawkins,

2006, P.31).

Quite a mouthful disparaging description of Yahweh, but let me begin by saying that Dawkins is a case of a biologist attempting to deal with a theological problem which is far from his grasp. As one man says, *"to expect to learn anything about important theological problems from Richard Dawkins…is like expecting to learn about medieval history from someone who had only read Robin Hood."* (Rodney Stark cited in Copan, 2011). It is one thing to factually and sincerely criticise something and quite another to put forward a self-made caricature and begin to ridicule and criticise it. Like the famous geneticist Francis Collins put it, *"Dawkins is a master of setting up a straw man, and then dismantling it with great relish"* (Collins, 2007, 164).

Consider this: in a small town of Zambia called Kabwe is a band of three dangerous criminals called *Mailoni Brothers*. They have managed to terrify an entire town, strangling, maiming and killing women, children and men in the most atrocious ways possible. Security agencies have failed time and again to arrest them. Meantime they take pride in the terror they create. A simple question: what would be the most reasonable thing to do with such ill-famed and hopeless personalities? Take them for counselling or kill them at any possible chance of spotting them? Well, the Zambian Police opted for the latter, and gunned down the criminals in the year 2014 and many people in the town and around the country celebrated the successful crack down. If some foreign so-called human rights activist was to visit the country and begin to vocally condemn the action as having been wrong because the criminals were killed before trial, such a person would be simply playing an academic exercise, so detached from the reality of the trauma experienced by many people who were victims of the three wild beasts. Like it or not, there are some situations where a person so completely loses his sobriety, becomes a terror and vermin in society, that he becomes no less different from a pile of cancer cells which are wreaking havoc to a human body.[1] This is exactly what influenced much of the Old Testament's harsh laws.

The Old Testament is a story of the fall of man and his subsequent moral decadence. Right from Genesis we are informed that *"the wickedness of man was great in the earth, and that every imagination of the thoughts of his heart was only evil continually"* (Gen. 6:5). It took Noah 120 years[2] to preach to wicked mankind to repent of the evil ways, but all his words fell on deaf ears[3]. In the time of Sodom and Gomorrah not even ten righteous people could be found there (Genesis chapters 18 and 19). With time various nations arose which were pagan and completely deluded to the extent of committing pagan atrocious acts of sacrificing infants into fire (Lev.18:21). This had to stop and there had to be a starting point[4]. Israel was the society chosen to begin living by God's prescribed moral

---

[1] It is in such cases that even the Bible would justify the legal sentencing of a death penalty on a criminal (Romans 13:3-4).

[2] The life-span of people during the antediluvian period was much longer than today's.

[3] Read Gen.6:3, cf. 5:32; 7:6 and 2 Peter 2:5

[4] In some cases where a tribe or nation was completely wicked and deluded, Israelites had to march through there and overthrow it. This had nothing to do with ethnic cleansing or genocide but suppressing forces which became dangerously morally repugnant – *"for the wickedness of these nations the Lord doth drive them out from before thee"*, said Yahweh (Deut.9:4).

standards. Note also that each time the nation fell in error, Yahweh did not play any partiality but let them get subdued by other (pagan) nations like Babylon. Clearly this is not a case of a racist or partial God.

For a people living at such a period of history when morals were at the lowest ebb, strict adherence and stiffest punishments were inevitable; a legislation of *"an eye for an eye"* has to be looked at in its historical context. It certainly won't apply in a modern setting where civilisation has made remarkable progress. But this was not so at the time of Moses when morality was so low that there had to be a detailed prescription of laws which even regulated people's diets, day of worship, manner of dressing, etcetera. This had to be done to continuously raise people's awareness of moral standards. And another important point to consider is that reading the Bible from the Old Testament through the New Testament presents God's wisdom in patiently and gradually improving man's morality. From the tough Mosaic legislation, written on tables of stone, which required various death penalties, to one written in the heart of man, requiring obedience through love for God and one's neighbour.

What we see in all this is that when a person reads the Bible with an aim to spot errors, it will be easy to isolate one verse *here* and another *there* and try to form a picture that satisfies an objective to disprove. With such an attitude and approach, the actual context and theme of scripture will either be missed or misconstrued. As has been discussed in the foregoing, it is simply not true that the killing of Canaanites was genocide or ethnic cleansing. Ethnic cleansing is fuelled by racial hatred ; one tribe or race hates another and then proceeds to destroying it. This wasn't clearly the case with Israel. And, as Dr Copan has extensively explained in his book, *Is God a Moral Monster?*, *"Dawkins isn't interested in accuracy; so he resorts to misleading rhetoric to sway the jury"*. He correctly notes that *"From the beginning, God told Abraham 'all the families of the earth' would be blessed through his offspring (Gen.12:3). We're not off to a very xenophobic start. Then we read many positive things about foreigners in the chapters that follow. Abraham...encountered just and fair minded foreign leaders among the Egyptians (Gen.12) and the Philistines (Gen.20) ...A 'mixed multitude' left with Israel from Egypt (Exod.12:38)...The gentile Rahab and her family joined Israel's ranks (Josh.6:23).* (Copan, 2011, p.163).

Now, all this discussion can only make sense if at all God exists. Many believe he exists, but, is there any proof or evidence for this?

2.
Proof or Evidence?
Getting our terms right

"The best way to show that a stick is crooked is not to argue about it or to spend time denouncing it, but to lay a straight stick alongside it"

Dwight L. Moody

Far from being an inhibitor or a contradiction to science, faith in God is actually what enthused early scientists to investigate nature and the cosmos. They perceived the regularity of laws exhibited in nature; the laws could be encoded and predicted using equations. These scientists found it reasonable to reckon that such precise order of the universe could only have proceeded from an all-powerful intelligent being. In the words of Johannes Kepler, *"The chief aim of all investigations of the external world should be to discover the rational order which has been imposed on it by God, and which he revealed to us in the language of mathematics"* (cited in Kline, 1980, p.31). The question to ask is, why did Kepler and many scientists of his day think that there was an unavoidable connection between "rational order" and "God"?

Science investigates phenomenon and patterns that emerge out of it. The discovered patterns or regularities enable scientists to *extrapolate*, and hence predict ideas and events. Thus, something *unknown* can be reckoned from evidence of the *known*. Now, while some extrapolations are testable and hence verifiable, others, no matter how rational, can only be affirmed in terms of their explanatory power. The latter is in no way unscientific because granted that there is enough experience (and therefore knowledge) about the *known*, the consciousness in man can have *faith* in the extrapolated *unknown*.

Consider this, cosmology explains the origin of the universe by the Big Bang model not because anyone ever saw or proved the Big Bang but because Hubble's evidence of the expansion of the universe entails that reversing the on-going expansion would bring all matter, energy and space into a single hot dense spot from which an explosion occurred which threw all matter into motion. Our faith, as far as present knowledge has revealed, can rest on that because *known* phenomena – the expansion of a balloon, for example – informs us that anything which is undergoing expansion must have previously been in a *contracted state*. Likewise if granted you were the first person to go to Mars and there you happened to find a writing engraved on a rock with the following English words and punctuation:

> *On eleventh September of the year 2001, the Twin Towers in New York City of the USA - destroyed in a terrorist attack.*

It would be unthinkable to suggest that because you never saw who wrote the words they were simply a result of wind continuously raging over the rock and eventually, letter by letter, producing the writing by chance. Your experiences on planet earth lets you know that the words exhibit *specified complexity*, that is so say, the engraving is not composed of arbitrary marks; the marks are very specified and are conveying a message. This characteristic of specified complexity has always been a product of intelligence and so it would be logical to conclude that the engraved writing was a product of intelligence. In this case you have faith in what you have believed not out of ignorance but through knowledge of the fact that such a meaningful series of symbols signifies an intelligent cause. Simply put, the symbols convey information and information in all its known forms never arises on its own (i.e. without cause). It always proceeds from *volition*, i.e., intelligent will.

Now, if it is plausible to infer the existence of something not by direct proof but rather evidence of manifest characteristics or features which are akin to other known phenomena, it is no exception that the specified complexity of information encoded in the DNA has an intelligent cause. This may not be proof but works as inference to the best explanation. It is for this reason that the *absence of proof* cannot always translate to *proof of absence*. And it therefore goes without saying that the absence of proof of God does not necessarily equate to his non-existence. There are many things we human beings believe through inference, and not proof of what *sight, taste, feel, smell* and *hearing* tells us.

Faith

Science in itself is silent on the subject of faith or atheism. Theism or atheism proceed from what perception one has from observed facts. Prudence requires that one adopts a view which is congruent with facts of evidence. A closer look will reveal that both the atheistic and theistic views are based on faith. However, both faiths cannot be right; one is either blind, incongruent or consistent with the facts of evidence. *Blind faith* merely believes for the sake of believing. This kind of faith turns out to be foolish when its adamancy persists in the face of contradicting evidence. There is also *irrational faith;* one which attempts to go some way to justify its premises but eventually turns out to be self-contradicting or failing to measure up with the available evidence. *Rational faith* is one which is consistent with the evidence.

Note that as long as we are dealing with things which cannot be proved, what we are relying on is the probability of the facts being true. That is what evidence refers to. Here is where we need to distinguish between the terms proof and evidence.

Proof or evidence?

It should be emphasised that far from meaning "*believing where there is no evidence*", true faith is "*the substance of things hoped for, the evidence of things not seen*" (Hebrews 11:1). But what is *evidence*? This word should not be confused with *proof*.

Proof is something that establishes an absolute fact, leaving no room for doubt. Before me is a computer screen which is *15 inches* long; you can prove the fact by simply measuring the length of the monitor. Proof can repeat and be verified. If all things could be as easily verifiable, we would have a problem-free world with no need for juries and magistrates. But we find ourselves in a world where certain things can only be ascertained by means of evidence or *inference to the best explanation*.

Evidence refers to a fact or situation that suggests something *might be* true. So while proof is about facts, evidence is expressed in terms of probability. When different facts seem to be consistent in support of an observation, we say there is *evidence*. Consider this: "*The universe began with a big bang*" – well, interesting statement. Hubble comes along and observes that galaxies are receding from each other, implying that space is expanding. This is more evidence for the Big Bang because expansion entails that all matter was once compacted in a single, hot, dense spot from which the explosion and subsequent expansion began[5]. But, there is more: next comes the Nobel Laureate, Arno Penzias, who discovers *microwave background radiation* in space and this is most likely the afterglow of the Big Bang. Now that's significant evidence!

Notice in the foregoing that we believe the Big Bang not because it can be proved but because of many interrelated facts that supply the evidence. In like manner many great scientists who were once atheists became believers on account of the much evidence that suggests a creator. Isaac Newton, arguably the greatest scientist who has ever lived, through knowledge about the precision of the laws that govern the universe, was a firm believer in God; in his most acclaimed book *Principia Mathematica,* he exclaimed that *"this Being governs all things, not as the soul of the world, but as Lord over all; and on account of his dominion he is wont to be called Lord God…This most beautiful system of the sun, planets, and comets, could only proceed from the counsel and dominion of an intelligent and powerful Being."* Francis Collins, leader of the Human Genome Project, used to be an atheist but was later converted to a theist on account that more evidence points to a creator God. In *Language of*

---

[5] I have come across naïve Christians who don't believe in the Big Bang because they suppose it contradicts Genesis 1:1 – *"In the Beginning God created the heaven and the earth."* Well, this is shear ignorance. Couldn't God have spoken, *"Let there be…"* and *"BANG!"* the universe went into explosion? It is sad to know that many ignorant Christians actually don't realize that the Big Bang theory is what has actually given serious headache to atheists who now have to grapple with the hard question of "who caused the Bang?" The Big Bang and God are not mutually exclusive explanations!

God he wrote: *"for me the experience of sequencing the human genome, and uncovering this most remarkable of all texts, was both a stunning scientific achievement and an occasion of worship"* (Collins, 2007, p.3). And the once most influential atheist Anthony Flew, after 50 years of atheism, gave the reason for his conversion; he said that the biologists' investigation of DNA *"has shown, by the almost unbelievable complexity of the arrangements which are needed to produce life, that intelligence must have been involved"* (quoted in Lennox, 2007, p.6).

These highly intelligent believers demonstrate that far from invoking God because of a gap in human knowledge, there are men of science and philosophers who have come to believe through reason. So belief in a creator-God is in no way diminished when we gain understanding of phenomenon which may have once been mystified due to ignorance. John Lennox's example of Henry Ford and the Ford vehicle makes good illustration:

Take a Ford motor car. It is conceivable that someone from a remote part of the world, who was seeing one for the first time and who knew nothing about modern engineering, might imagine that there is a god (Mr. Ford) inside the engine, making it go. He might further imagine that when the engine ran sweetly it was because Mr. Ford inside the engine liked him, and when it refused to go it was because Mr. Ford did not like him. Of course, if he were subsequently to study engineering and take the engine to pieces, he would discover that there is no Mr. Ford inside it….But if he then decided that his understanding of the principles of how the engine works made it impossible to believe in the existence of a Mr. Ford who designed the engine in the first place, this would be patently false…had there never been a Mr. Ford to design the mechanisms, none would exist for him to understand (Lennox, 2009, p.45).

Clearly God does not disappear when science explains, but rather seems even more probable. It is also important to note that an explanation of science can never be ultimate in itself because, as the Oxford philosopher, Richard Swinburne, explains, the very capacity of science to explain also requires explanation (Swinburne, 1996). Prudence requires us to follow the advice of Socrates – *examining the evidence and seeing where it leads!*

3.
Origin of the Universe
Which way does the evidence lead - God or Chance?

"It is said that an argument is what convinces reasonable men and a proof is what it takes to convince even an unreasonable man. With the proof now in place, cosmologists can no longer hide behind the possibility of a past-eternal universe. There is no escape, they have to face the problem of a cosmic beginning"

Alexander Vilenkin

At one time it was very easy for atheists to explain that the universe has been eternal. This was a simple way of avoiding the problematic question of '*What* brought about "*the beginning*"?' This would be a difficult question to handle because all our experiences in the world follow the law of cause and effect – nothing causes itself. However, the idea of an eternal universe was turned on its head when it was discovered that the universe is actually undergoing expansion.

If space is expanding then it naturally follows that at one time all matter began from a single hot dense spot from which an explosion occurred throwing everything into motion. Stronger evidence of the Big Bang emerged with the discovery of microwave background radiation. This fact of a cosmic beginning was further augmented by three leading cosmologists Borde, Guth and Vilenkin (2003) who demonstrated that a universe which is perpetually expanding cannot have an infinite past but must have a finite space boundary – "*almost all causal geodesics, when extended to the past of an arbitrary point, reach the boundary of the inflating region of space-time in a finite proper time*" (p.3). In *Many Worlds in One* Vilenkin (2006) is blunter: "*With the proof now in place, cosmologists can no longer hide behind the possibility of a past eternal universe. There is no escape, they have to face the problem of a cosmic beginning*" (p.176). But, why is it a "problem"? The *Kalam*[6] cosmological argument explains.

---

[6] *Kalam* is an Arabic word for "Theology." The proponents of this argument were Muslim scholars who lived in medieval times.

The *Kalam* Cosmological Argument

The *Kalam* cosmological argument explains why a cosmic beginning gives atheism a hard time. This argument rests on the premise of the following simple logic:

- Anything that begins to exist has a cause
- The universe began to exist.
- Therefore, the universe has a cause.

i) *Anything that begins to exist has a cause*

Clearly, one can't argue against this first premise. We live in a world where nothing has ever caused itself. As in the words of Craig (2010), *"To suggest that things could just pop into being uncaused out of nothing is literally worse than magic."* So, *premise 1* does not require a sophisticated proof; our everyday life experiences readily provide experiments of *effect* always following *cause*. However, there are *things* which exist without cause but are simply there for necessity; numbers or sets for example are not caused but exist necessarily. Of importance to note here is that things that are caused to exist can never exist necessarily; they exist by reason of a cause (Craig, 2010)[7].

Professor William Lane Craig is credited for having popularised the argument and is the one who has named the argument Kalam, in honour of its proponents.

[7] Some scientists have tried to explain that something can come up from nothing without cause. They do this using the concept of *virtual particles* arising 'uncaused' in 'empty' space. But the absurdity of this assumption soon emerges when we realise that empty space is not really empty when we consider the charge of energy it contains. Clearly energy is *something* and not *nothing*!

ii) *The universe began to exist*

This premise is supported by the evidence of philosophical reasoning as well as science.

Philosophically, it is illogical for an infinity number of past events to exist. The very word "event" denotes occurrence and hence cause. An infinite regress of events is absurd. It is simply unavoidable for past events to be finite. So, if the universe evolved through various past events which rewind back to the Big Bang, no explanation can rule out a termination of those events, not even the speculation of the multiverse can. Whatever events can be explained, whether at the macro or quantum level, an initial start-up point is inevitable as long as it concerns things that are caused.

The scientific evidence of this premise is based on the expansion of the universe as was earlier explained. The universe is expanding and this clearly implies that rewinding the expansion process will have all matter, space and energy wind up in a condensed initial start-up (known as the *singularity*) from which the Big Bang occurred. The strong evidence of the afterglow of the explosion (i.e. the *microwave background radiation*) simply establishes the fact that the universe began to exist and it logically follows that

---

Now, is it even conceivable that one day there will be an experiment to demonstrate that something can come out of nothing? The experimenter will need, among other tools, specimen and apparatus, *nothing*. Well, it is possible that exaggerated optimists of scientism exist; they may dare to go on a voyage to discover the space *filled* with nothing. I would bid them bon voyage, but am afraid it won't be long in their sail of the unexplored waters before they get shipwrecked by the laughter of the *gods*.

this existence was caused and hence the universe is not a necessary being like numbers.

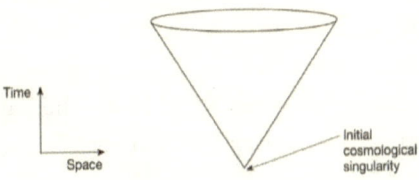

Figure 3.1: Geometrical Representation of Standard Model Space-Time. Space and time begin at the initial cosmological singularity, before which literally nothing exists[8]

Interestingly with the presentation of models that gave convincing evidence of a universe that must have begun to exist, research into alternative models followed, all skewed towards avoiding an initial beginning that smacks of divine cause. Models included the steady-state model, oscillating model and quantum gravity models, among others.

Steady-state model

This model was proposed by Fred Hoyle. He acknowledged that truly indeed matter in the universe is moving away from each other and there is clearly an expansion of space occurring. However, reversing the expansion process does not take us to a violent beginning point where matter is heavily condensed into a singularity from which a bang of explosion occurs. Hoyle proposed a *steady-state model* in which there is no violent beginning of the universe but a steady continuous process of new

---

[8] Source (for this and all the following models): Astrophysics and Space Science, 1999.

matter emerging in empty space. He explained that as matter moves away from each other through an expansion process, empty spaces are left behind which gradually gets filled with newly formed matter. This process is continuous and therefore implies that the density of the universe is constant (see fig.3.2). The theory claimed that the universe is *homogeneous* (i.e. looks the same in every place) and is also *isotropic* (i.e. looking the same in every direction).

The Steady State model was clearly just another version of an eternal universe. It was later laid to rest as evidence of a non-uniform history of the universe emerged. The last nail in its coffin was the discovery of *microwave background radiation*. The temperature of the radiation corresponded with the temperature that should have been left after the Big Bang - $3°C$ above *absolute zero*. This became the strongest evidence of the Big Bang.

Oscillating model

Assuming that the universe is not homogenous and isotropic, and that if the energy that causes the universe to expand is weaker than the gravitational pull of celestial matter, then the expansion would at a certain point halt and begin to reverse into contraction. Now because the universe is not the same in every place and direction, there will be room for the contracting matter to simply pass each other by (i.e. without coalescing at a certain point) so that the new direction by contraction would appear to go into a new expansion phase (see fig. 3.2).

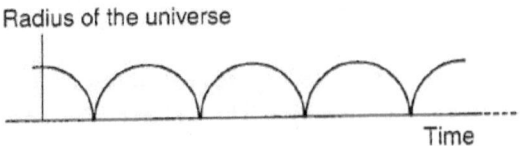

Fig.3.2: Oscillating Model. Each expansion phase is preceded and succeeded by a contraction phase, so that the universe in concertina-like fashion exists beginninglessly and endlessly.

This speculative model never survived for long as Penrose and Hawking (cited in Craig, 2010) disclosed that under generalised conditions an initial cosmological singularity is inevitable, even for homogenous and non-isotropic universes. Craig notes:

First, there are no known physics which would cause a collapsing universe to bounce back to a new expansion. Second, the observational evidence indicates that the mean mass density of the universe is insufficient to generate enough gravitational attraction to halt and reverse the expansion. Third, since entropy is conserved from cycle to cycle in such a model, which has the effect of generating larger and longer oscillations with each successive cycle, the thermodynamic properties of an Oscillating Model imply the very beginning its proponents sought to avoid (See fig.3.3).

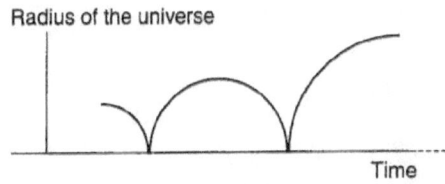

Fig. 3.3: Oscillating Model with Entropy Increase. Due to the conservation of entropy each successive oscillation has a larger radius and longer expansion time.

So many models can be discussed but let me wind up with

a current one by Hartle and Hawking which is explained using the concept of quantum gravity.[9] Hartle and Hawking explain that the concept of *time* as we know it cannot apply beyond the beginning. The Hartle-Hawking model therefore makes use of imaginary numbers to represent time. By doing this the singularity is eliminated so that although the past is finite, it has no beginning point. Like the earth which is round and for which if you were to run around it you won't come to a boundary point or fall off an edge, so is the universe. Thus the standard conical-hyper-surface is transformed into a curved hyper-surface (see fig.3.4 and compare with fig. 3.1).

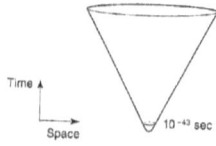

Fig. 3.4: Quantum Gravity Model.

Stephen Hawking further claimed that the universe caused itself to exist by the law of gravity. With all due respect, I must say that this statement was too illogical to have been made by a man of his impeccable genius. But, shall we believe such irrationality simply because the proponent enjoys immense prestige? Oxford mathematician, John Lennox speaks:

The simple law of arithmetic by itself, 1+1=2, never brought

---

[9] It is not the purpose of this author to discuss technical details of these models. For a detailed exposition on the models visit www.reasonable.org.

anything into being...If I put £1000 into the bank, and later another £1000, the laws of arithmetic will rationally explain how it is that I now have £2000 in the bank. But if I never put any money into the bank myself, and simply leave it to the laws of arithmetic to bring money into being in my bank account, I shall remain permanently bankrupt (Lennox, 2011, p.41).

The Oxford mathematician admonishes: *"What this all goes to show is that nonsense remains nonsense, even when talked about by world-famous scientists"* and *"immense prestige and authority does not compensate for faulty logic"* (p.32).

iii) *The universe has a cause*

To-date, all the different models have failed to convincingly challenge the standard model. Professor Craig, after analyzing all the different models and discussing the issue of existence by necessity and by contingency, has beautifully presented the following logic:

1. *Whatever exists has a reason for its existence, either in the necessity of its own nature or in an external ground.*

2. *Whatever begins to exist is not necessary in its existence.*

3. *If the universe has an external ground of its existence, then there exists a Personal Creator of the universe, who, sans the universe, is timeless, spaceless, beginningless, changeless, necessary, uncaused, and enormously powerful.*

4. *The universe began to exist.*

   From (2) and (4) it follows that:

5. *Therefore, the universe is not necessary in its existence.*

   From (1) and (5) it follows further that:

6. *Therefore, the universe has an external ground of its existence.*

   From (3) and (6) we can conclude that:

7. *Therefore, there exists a Personal Creator of the universe, who, sans the universe, is timeless, spaceless, beginningless, changeless, necessary, uncaused, and enormously powerful.*

   And this, as Thomas Aquinas laconically remarked, is what everybody means by God.

I can't agree more. And, when I think of all the failed attempts to challenge the cosmic beginning, a verse comes to mind:

Because that which may be known of God is manifest in them; for God hath showed it unto them. For the invisible things of him from the creation of the world are clearly seen, being understood by the things that are made, even his eternal power and Godhead; so that they are without excuse (Romans 1:19-20).

Now, after "*the beginning*" more mind-boggling marvels and evidence of design emerge. Scientists have been able to model events that occurred in the time just at the time of the Big Bang. What is astonishing is that if these early conditions were altered in any slight way, life as we know it would never have existed. The universe is amazingly fine-tuned!

4.
Fine-tuned Universe
How come life?

"What are we to make of this?...a common response was to shrug the matter aside with the comment, 'The value it has is the value it has, and if it had been different we wouldn't be here to worry about it'...[but] *the fact that the value of the strong and electromagnetic forces in atomic nuclei are 'just right' for life ...cries out for explanation"*

Paul Davies

An atheist-scientist was once asked a question: What do you consider to be the strongest argument for the existence of God? *"The fine-tuning of the universe"*, he responded. Eminent atheist physicist Stephen Hawking also once remarked that *"the odds against a universe like ours emerging out of something like the Big Bang are enormous. I think there are clearly religious implications."*

So, what is "fine-tuning" and why is it such a strong argument for the existence of a cosmic designer?

Consider this: let's go back in time and imagine that you have an opportunity to make adjustments to the various initial conditions at the beginning of the universe. How far do you think you can go in making changes (to the strength of gravity, the strong and weak nuclear forces, e.t.c) before spoiling things and causing a situation that would hamper the development of life?

The universe, right from the moment of the Big Bang, was so finely tuned that making any simple adjustments to its physical constants would yield completely different scenarios which would inhibit life. This has puzzled great minds for a long time. Fine-tuning (commonly known as the *anthropic principle*) refers to this delicate precision of the fundamental constants of the universe that made life possible to exist. Examples include the precision of the force of gravity, expansion of the universe, and the efficiency of nuclear fusion.

Gravity

This is perhaps the most mysterious and perplexing force in the universe. It is the glue that holds everything in the universe. It is omnipresent and omnipotent - "*No substance, no kind of particle, not even light itself escapes its grasp*", writes Rees (2000). Without gravity the force within the interior of stars will succeed in pushing outwards and the universe will explode. After the Big Bang, as the temperature began to cool down, the force of gravity caused matter to begin to coalesce into galaxies, planets and other celestial bodies. It is important to note here that the strength of this force isn't arbitrary. It is so precise that any alteration will be disastrous.

Strangely the strength of gravity (denoted by the symbol N) is actually a weak force; it is 1,000,000, 000, 000, 000, 000, 000, 000, 000, 000, 000, 000 (that is, 10^{36}) times weaker than the force that holds protons together. But, removing only six zeros from this large number would have stunted the formation of planets – "*Paradoxically, the weaker gravity is…the grander and more complex can be its consequences. We have no theory that tells us the value of N. All we know is that nothing as complex as humankind could have emerged if N were much less than 1,000,000, 000, 000, 000, 000, 000, 000, 000, 000, 000*" (Rees, 2000, p.31). And if it were stronger? Galaxies would have formed very quickly but would not be as large as they are; stars would be so close and densely packed bringing about planets with unstable orbits. The precision of gravity is tuned on a razor edge!

But, someone might say, such strange things could happen by chance. Yes, but not when one such thing leads

Fine-tuned expansion of the universe

In the 1920s the famous American astronomer made a ground breaking discovery when he observed that galaxies are flying apart from each other. This meant that space is expanding, and if we were to look back in time of this process, we can rightly predict that there was a moment before matter began to move away from each other; the matter was very close together, and going further back in time we arrive at a point where all matter and energy were heavily condensed in a single hot spot from which – "BANG!" – the explosion occurred, throwing all matter into motion. The energy of the expansion was so immense that to-date galaxies are still flying apart from each other.

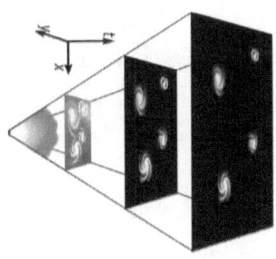

Fig. 4.1: *Illustration of the Big Bang and subsequent expansion*[10].

However, note that when an expansion is occurring it

---

[10] Image: Wikimedia

wrestles against the force of gravity. It is for this reason that when an explosion occurs flying debris is later overcome by gravity and hence eventually collapses to the ground. If a bomb or an explosive were to have a great propulsion of energy, so strong that it perpetually exceeded the pull of gravity, its debris will fly away from the earth. Of special note here is that there is a point at which expansion energy either exceeds gravity or is overcome by it - the *critical point*. It ofcourse follows that the further the measure and speed of energy beyond the critical point, the longer the duration of expansion. Now, how far can an expansion which results from an explosion whose energy is not much beyond the point at which a collapse could occur (i.e., the expansion is occurring near the critical rate) go? It is here that experts tell us of a paradox concerning the manner in which the expansion of the universe occurred in the beginning: The universe started out with "*so nearly the critical rate of expansion that separates models that recollapse from those that go on expanding forever, that even now, ten thousand million years later, it is still expanding at nearly the critical rate*" writes Stephen Hawking in *A Brief History of Time*. Hawking further noted that: "*If the rate of expansion one second after the Big Bang had been smaller by even one part in a 100 million million, the universe would have recollapsed before it ever reached its present size*" (1998, p.138). Interestingly again, if the rate of expansion had been slightly greater, by even one part in a million, stars and planets could not have been able to form!

Nuclear efficiency

Things are made up of very tiny particles called *atoms*.

These atoms are in turn made up of smaller particles, the *nucleus* and *electrons*. The nucleus stands in the central area of an atom and consists of *protons* and *neutrons*. Protons in turn consist of smaller particles called *quarks*.

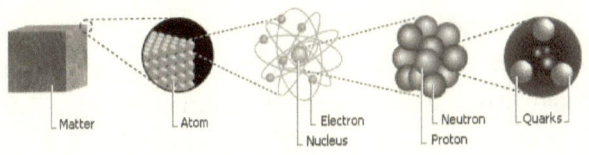

Fig.4.2: Fundamental particles of matter[11]

Protons are electrical particles with a positive charge whilst neutrons are particles which do not carry an electrical charge, i.e. they are neutral. The nucleus (consisting of protons and neutrons) is surrounded by negatively charged particles called electrons. Note that particles with like-charge repel and those with opposite charges attract. So a proton will repel a proton, and likewise an electron will repel an electron. Now here is a little puzzle - if particles of like-charge repel, how do protons in the nucleus of an atom hold together firmly? The answer lies in the fact that quarks are firmly held together by what is known as *the strong nuclear force* and it is the remnant of this inter-quark force that holds protons together, overcoming the electrical repulsions between them. However, when nuclei of different atoms collide at high temperature and speed, protons coming from different atoms can come so close

---

[11] Image: Microsoft Encarta

that their repulsion gets overcome by the strong nuclear force and hence causing them to fuse, forming a new heavier element. So for example, when an element which has a nucleus with a mass of two (a type of hydrogen atom called *deuterium*) fuses with another with a mass of three (a type of hydrogen atom called *tritium*), a new element called *helium* is formed. Of special interest to note here is that if the nuclear force were weaker only hydrogen could have formed in the universe (closing off the formation of other essential elements). Again if it were slightly stronger all hydrogen in the early universe would have converted into helium and hence leaving no possibility for the formation of water an essential ingredient for life (See **Appendix I**).

One more marvel about nuclear efficiency concerns the formation of *Carbon*, a very crucial element for the existence of life on earth. It consists of 6 protons and 6 neutrons. Carbon is formed by combining three nuclei of helium. However, there is very little chance of the three coming together simultaneously. Thus the formation of Carbon takes place through an intermediary stage of first two helium nuclei combining and forming *beryllium* (consisting of four protons plus four neutrons) before combining with a third helium to produce carbon. *Problem*: The beryllium nucleus is so unstable that it would quickly decay before a chance of a helium nucleus coming to combine with it. *Question:* If beryllium is so unstable like that, how does carbon form? Fred Hoyle solved the riddle: A carbon nucleus has a peculiar feature – a 'resonance' with a particular energy that enables beryllium to grab another helium nucleus in the brief interval of time before it decays: *"The frequency with which the three helium nuclei vibrate as they come together exactly matches one of the natural frequencies of a vibrating carbon nucleus",* explains Goswami (2012). The marvel is: *"How would the three helium nuclei know how to dance one of a select few dances that six protons and six neutrons of the carbon nucleus can perform?"* (Goswami, 2012). Note that if the nuclear force were slightly more attractive no carbon would exist – it would all convert to oxygen! Fred Hoyle had this to say concerning the marvel of carbon:

> *Would you not say to yourself, Some super-calculating intellect must have designed the properties of the carbon atom, otherwise the chance of my finding such an atom through the blind forces of nature would be utterly minuscule. A common sense interpretation of the facts suggests that a superintellect has monkeyed with physics, as well as with chemistry and biology, and that there are no blind forces worth speaking about in nature. The numbers one calculates from the facts seem to me so overwhelming as to put this conclusion almost beyond question* (Hoyle, 1981).

What do we make of all this?

The fine-tuning of the fundamental constants of the universe has always astounded scientists. Physicist Paul Davies asks: *"What are we to make of this?...a common response was to shrug the matter aside with the comment, 'The value it has is the value it has, and if it had been different we wouldn't be here to worry about it'...*[but] *the fact that the value of the strong and electromagnetic forces in atomic nuclei are 'just right' for life ...cries out for explanation."* (Davies, 2006, p.157).

Francis Collins summarises three options which are often given by different scientists of what could be the explanation of the riddle of fine-tuning:

i) *Luck:* There is only one universe and we just happen to be *here* by luck. The right characteristics that gave rise to life were simply by chance.

ii) *Multiverse:* There could be an infinite number of universes, and each of these could have different physical constants and laws. We happen to exist in a universe where the physical laws permit life.

OR

iii) *God:* There is only one universe and it's precise physical constants were purposely 'tuned' by God.

Evaluating the options

Luck: This is extremely improbable and cannot be taken seriously.

Multiverse: Is it not simple logic to reckon that much as multiplying the number of universes is a clever trick of trying to underplay fine-tuning and hence rule out the possibility of God, it in itself does not necessarily rule out the fact that ultimately there had to be a beginning of them and this would take us back to the same problem of a cosmic beginning? The multiverse concept is only a good mental gymnastic as it is highly speculative, and in the words of Manson (2003), *"the last resort for the desperate atheist"* (p.18). A number of scientists and philosophers have not taken the speculation seriously: Said Paul Davies, *"The disadvantage of the multiverse theory is that it invokes an overabundance of entities, most of which could never be observed, even in principle... the theory is very hard to test*" (Davies, 2006, p.298). And Polkinghorne; "*There is no purely scientific reason to believe in an ensemble of universes...these other worlds are unknowable by us*" (Cited in Lennox, 2009). Sir Martin Rees described it as being "*highly speculative*" (Rees, 2003, p.164), and Richard Swineburne, University of Oxford philosopher, rubbishes it as "*the height of irrationality*" (Swineburne, 1995).

God: Our third option can be compared with (*ii*) above through an illustration of a parable used by Leslie:

A firing squad of highly trained marksmen aim their rifles on an individual tied to a pole. The order is given and they shoot and it happens that they all miss the target and the condemned fellow is unscathed.

What would be the likely explanation of this event? A Multiverse-equivalent explanation can claim that there may have been a thousand executions carried out on that same day and they all missed the target because even the best marksmen occasionally miss. The God equivalent explanation will state that something directed is going on – *the missing off the target was intentional!*

Which explanation makes sense? Clearly the God-explanation. However, there could always be someone with too strong unbelief in anything that suggests the existence of God. There was such a renowned atheist, a 'Dawkins' of the bygone era, who won public fame in his arguments against the existence of God. But the unexpected happened. After having been a prominent atheist for over 50 years, he ended up a believer of God. What convinced him? The miracle molecule of DNA whose operations works in accordance with its precisely set information! Information by definition can never arise on its own. As far as all experiences in this world have shown, information only emerges from intelligence. The crucial question: who or what programmed the information in the DNA?

5.
The Information Evidence
Whence came information?

"I get down and write in one square my best set of equations for the universe, and you get down and write yours...we have our magic wand and give the command to those equations to put on wings and fly. None of them will fly. Yet there is some magic in this universe of ours, so that with the birds and the flowers and the trees and sky it flies. What compelling feature about the equations that are behind the universe is there that makes them put on wings and fly?"

John Wheeler

Atheists (or *materialists*, to be more specific) believe that everything we see around us can be explained in terms of matter. That is to say, there is no such a thing as a spiritual dimension where God, angels, or demons exist. Materialism further posits that even the difficult problem of explaining consciousness will one day have a naturalistic explanation. Indeed many materialists already explain that thoughts occur in the brain and are hence reducible to brain cells which in turn are composed of molecules, which are made up of atoms. This approach of perceiving complex matter in terms of simpler units is known as *reductionism*. Some reductionists have presumed that any alleged testimony of experiencing a spiritual vision or divine power is simply a result of arousal of some particles in the brain. Francis Crick is quite emphatic on this notion: *"you, your joys and your sorrows, your memories and ambitions, your sense of personal identity and free will, are in fact no more than the behavior of a vast assembly of nerve cells and their associated molecules"* (1994, p.3). It follows from this worldview that human rationality – thoughts, consciousness or intentionality - arose from the motion of non-guided irrational atoms. So, a rock, frog and human being are all mere chunks of matter in different stages of evolution!

Notice that a materialistic worldview is logically incompatible with the existence of objective morality. When a chunk of matter called a rock falls on another rock, we do not describe the occurrence as an *assault* or act of *murder*. Likewise, when another kind of cluster of atoms called a rooster forces itself onto a hen, the action is not *rape* but mere *copulation*. It is said that this is applicable to human beings (who are taken to be just another kind of animal species), except that their consciousness has made them construct moral words of *murder, rape, fornication* etcetera. This morality is not objective truth but a mere psychological trick which works to flourish the existence of the sentient creatures. Thus, in this atheistic worldview, good and evil do not really exist – "*In a universe of electrons and selfish genes, blind physical forces and genetic replication, some people are going to get hurt, other people are going to get lucky, and you won't find any rhyme or reason in it, nor any justice. The universe that we observe has precisely the properties we should expect if there is, at bottom, no design, no purpose, no evil, no good, nothing but pitiless indifference*" (Dawkins, 1995).

One other argument often used by materialists to argue against the existence of souls, spirits or the supernatural in general arose as a response to Rene Descartes'[12] philosophy which explained that minds are made up of an immaterial substance which does not occupy a place in space. This idea was objected by an explanation which stated that physical matter can only influence other physical matter; contact is required for one thing to move another; so, how can an immaterial soul have a casual effect on material substance?

---

[12] Rene Descartes (1596-1650) was a French scientist, mathematician and philosopher. He is known as the father of modern philosophy.

The poverty of materialism

First thing: if what materialists explain is *true* then it is not *true* because *truth* does not exist in their worldview. Moreover, *"If my mental processes are determined wholly by the motions of atoms in my brain, I have no reason to suppose that my beliefs are true...and hence I have no reason for supposing my brain to be composed of atoms"* (Huldane cited in Lewis, 1974). Think of it, why should one chunk of matter (in whose worldview truth doesn't exist) bother to go to great lengths trying to convince the other chunks of matter about the truth? This is an attempt to *prove* that *proofs* don't exist; presenting *an argument* that *arguments* don't exist (Lewis, 1974). Quite ridiculous absurdities! And if the human mind is no more than a product of neuro-circuits, how come we reason, not only within ourselves but with each other, and we also have intentions. Borrowing words of Polkinghorne (1986), I ask: if thoughts are merely electro-mechanical neural events, how can two such events confront each other in rational discourse? Free will is clearly evident in our everyday life experiences. A world which is completely material can never have a place for free will or intentionality. And isn't the denial of the existence of free will evidence in itself that free will exists or the person wouldn't have made the choice of that argument?

The information challenge

Among the greatest discoveries in biology was the information code that guides the development and growth of organisms. The information processing mechanism occurs in a molecule called the DNA. Before explaining what this entails it is essential to know what *information* means.[13]

I am right now using Microsoft Word to type these words. When I press the letter "A" on the keyboard, an electrical pulse train representing the symbol is initiated which is later conveyed to my eyes through the monitor screen. This simple process is not as simple as it may seem to a computer user. Underlying the whole process are coded instructions which perform the operations. The codes are what constitutes computer software. So, the software instructions are what *informs* a computer system to perform tasks. Of great importance to note here is that much as the computer appears to be so smart in the way of executing tasks, there are actually underlying instructions that provide the information of what needs to be done.

If a computer user would look at how the actual code of software looks, he or she may never make out what it means. The code used to instruct the machine consists of binary digits of *ones* and *zeroes*. Incomprehensible as this may appear to an untrained eye, the binary codes are not haphazard; their particular combinations provide specific information for the execution of tasks.

Now, it is a simple matter of fact that such precise information, required for the orderly execution of tasks, had to be programmed by human intelligence. To use an illustration similar to Lennox's Ford car example, granted if one was a stranger from an uncivilized remote place, and was given an *HP* Personal Computer, it would certainly be a marvel for him to experience the use of the sophisticated

---

[13] Professor Werner Gitt has devoted an entire excellent book on this subject (Gitt, 2000).

capabilities of the machine. Let us suppose that in his ignorance he comes to believe that there is a god - Mr Hewlettt Packard - inside the circuitry of the computer that enables it to perform the unfathomable feats. He then supposes that this god is very gracious whenever the *Google* search engine works very fast to answer his 'prayers', and when *"Unable to connect to the Internet"* is displayed on the browser, he takes this to mean that god is angry and unwilling to answer his queries. Well, the person later learns that there is actually no god in the computer but only a series of software instructions. Would it be wise for the stranger to reckon that since he now understands how the mechanism works, Mr Hewlett Packard or programmers do not actually exist? Surely that would be a foolish assumption because understanding a mechanism does not rule out the existence of an agency which made the mechanism. The information processing mechanism in the computer is evidence that there was intelligence involved for its capabilities. So, instead of dismissing the existence of programmers, new understanding of how coded instructions work should actually cause more marvel at the great intelligence that enables the computer to work without human's direct intervention. Well, the interesting thing is that the discovery of the DNA revealed astounding phenomenon of instruction processing that works in the same way as that of computer code!

How DNA works

The letters DNA are an acronym for Deoxyribonucleic Acid. It is a nucleic acid molecule which is found in all living things (except some viruses) and comes in form of a twisted double strand double helix. DNA stores genetic information. Now just as information used by human beings to communicate and give instructions to each other is made by arranging letters of the alphabet in different ways, the DNA's alphabet consists of only four chemical bases, namely *adenine* (denoted by symbol A), *Guanine* denoted by symbol (G), *Cytosine* (C), and *Thymine* (T) (See fig.5.1 below).

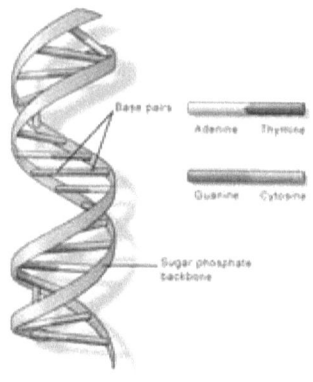

Fig.5.1 *Illustration of components that constitute DNA*[14].

The order and sequencing of these bases convey information for the creation of different proteins. So, for example the combination of *G, A* and *C – GAC –* is a genetic 'word' or code for the formation of *leucine,* an essential amino acid. This is just one among the many complex information processing capabilities of the DNA.

---

[14] Image: US National Library of Medicine

This information processing unit is what guides the development of features of a living thing. It is what defines an acorn to be an acorn and not a mango; a monkey to be what it is and not a human being.

It is important here to note that although the chemical bases of DNA are a physical medium, the information they convey is non-physical!

Here is an illustration:

Read this question:
What do you see after this question?
X

First you were instructed to do something, that is, "read this question". Your mind recognized the alphabetical letters and their combinations and from this it interpreted the meaning. That meaning in turn caused your eyes to move focus to the next line of words showing the question, and then you were further made to respond in your mind that it is letter X that you saw. Now please note the following facts about this process:

i) The words you read are made up of ink which is pressed onto the paper. Both the ink and paper are media which have been used to convey the information.

ii) The meaning of the text is not in the ink or the paper. If this were so, a person who only understands Chinese and wishes to understand the text can decipher the meaning by simply examining the chemical contents of the ink and paper. But this will be a foolish exercise as the meaning of a message can never be an intrinsic part of the code used for its composition.

iii) Just as the instructions were written and communicated using the media of ink and paper but yet with the meaning not intrinsic in the media, so the interpretation of the meaning of the text occurs through the avenue of the brain structure but is itself not a part of the cell structure of the brain cells. Cells are reducible to molecules and atoms and these particles cannot interpret anything.

All this trickles down to the fact that information is not material – it is "*neither a physical nor a chemical principle like energy and matter, even though the latter are required as carriers*" (Peil cited in Gitt, 2000). It is for this reason that information, in a semantic sense, cannot be quantified the way we are able to quantify matter.

Another very important thing to know about information is that in all its known forms it is always caused by volition (i.e. someone's will). Information never arises on its own; it is always caused by an intention or a willful act. Here is where there is a flaw in the concept that a physical material can only be caused to do something by another physical material: information is non-material but it (through thoughts or volition) causes matter to do something. Information actually comprises the nonmaterial foundation for all technological systems and for all works of art – "*All technological systems as well as all constructed objects, from pins to works of art, have been produced by means of information. None of these artifacts came into existence through some form of self-organization of matter, but all of them were preceded by establishing the required information*" (Gitt, 2000, p.49).

With all the facts presented in the foregoing, if we are to go back further to the point which evolutionists say life emerged from non-living matter, a more serious problem arises: what sort of information guided the transition from inert matter to living organisms? Now, an overreach to the moment when matter first burst into existence and began to evolve into a fine-tuned universe: what caused or guided this development? "Laws!" cosmologists would say. Well then, that would mean laws are information and therefore would need to be preceded by will (volition). So, whose will put them there? What gives *life* to these laws? The theoretical physicist John Wheeler wondered:

> *I get down and write in one square my best set of equations for the universe, and you get down and write yours...we have our magic wand and give the command to those equations to put on wings and fly. None of them will fly. Yet there is some magic in this universe of ours, so that with the birds and the flowers and the trees and sky it flies. What compelling feature about the equations that are behind the universe is there that makes them put on wings and fly?* (cited in Goswami, 2012).

Wheeler's words echo an ancient Hebrew scripture: "*Have you grasped the celestial laws? Could you make their writ run on the earth?*" (Job 38:33).[15] Like it or not, one fact stands – in the beginning was information which proceeded from a transcendent, non-material, and timeless being. The record

---

[15] The Jerusalem Bible

of Saint John is correct, "In *the beginning was the Word, and the Word was with God, and the Word was God… All things were made by him; and without him was not anything made that was made. In him was life*" (John 1:1-4).

6.
The Moral Argument
How come the moral compass?

"A man says about something he wants to do; 'it can't be wrong because it doesn't do anyone else any harm'…He is thinking it does not matter what his ship is like inside provided that he does not run into the next ship…He quite understands that he must not damage the other ships in the convoy, but he honestly thinks that what he does to his own ship is simply his own business. But does it not make a difference whether his ship is his own property or not?"

C.S. Lewis

"A city in Germany has legalised public nudity through the introduction of six designated nudist zones - one of which is situated in a main tourist spot, just minutes away from Munich's main square. State wide laws prohibiting nude sunbathing expired in Bavaria in 2013 and public nudity has been a hot topic of debate ever since" – read the news on *The Independent* and a plethora of other news websites (15th April, 2014).

This is what began as a small club – *"Free the Body Culture"* – in 1898 in Essen, but is now increasingly gaining public attention. This is not the only strange campaign around; there are now different outcries in the West for more of such freedoms even when it seems immorality is already on its peak. Today a campaign for women to be freely exposing their nipples in public is becoming a hot issue. Advocates have protested that allowing men to freely expose their chests whilst prohibiting women folk (in public or through social media) is an injustice. Oh! Just how has morality gone through such a rapid trajectory where people now fail to distinguish between a male's chest from a female's? We certainly are living in a time when common sense is no longer common, and the trouble is, when common sense leaves, nonsense remains. Explaining the difference between a male's chest and a female's should not require a biologist; one just needs to look around for a commercial bill-board advert featuring something like engineering equipment but which is superimposed with a figure of a sexy-looking figure of a woman, with either breasts or thighs made somewhat conspicuous to lure attention. Is it so hard to think why the legs or chest of a man won't do the same job?

A human being is a creature of free will and always craves for more freedom. No one knows the magnitude of stupidity that would be expressed if freedom were granted without restriction. Let us beware that freedom is not truly freedom if it is not defined or restricted by laws; 'Undefined' freedom will soon or later imprison a people with all manner of social ills. For freedom to co-exist with peace restrictions are inevitable. Does not beauty exist because a river has a specified course of flow and a rose has a particular blend and texture of color and fragrance? On the converse putridity would be the natural result of *everything* wanting to be *anything* and in *everyplace*. It has been said that nature always favors disorder. This is known as *entropy*. When the worse comes to the worst *social entropy* eventually culminates into *anomie,* a condition of social instability caused by the abandonment of moral standards. But…

Whose standard?

If morality entails following acceptable standards, the question that arises is, Who defined these standards? In other words, for those who object nudity, who said nudity is bad? Clearly it was only legalized in certain places in Germany and not everywhere else because there are some people who consider it unacceptable. But by what objective standard is it unacceptable? To say, "because the constitution or laws of the country does not allow it" is not answering the question as it only makes the moral issue a subjective piece of legislation which may not be to the taste of other people. Moreover, proponents of nudity would ask about what harm or injury the act has on other people when nudists are just freely expressing a natural 'non-harmful' state. Furthermore it has been said that as long as something – sex, nudity, pornography – has no bearing on other people, but that the individual or consenting persons are happy with it, then freedom of expression should be granted. Proponents of this view on morality restrict the subject to be only concerned with "*fair play and harmony between individuals*" and it is here where they get it wrong.

Morality is not only a matter of what you do and how it affects another person. Lewis (1944) correctly listed two other things that constitute morality: the harmonizing of things inside each individual and the general purpose of human life. He used an analogy of a fleet of ships on a sea to illustrate this:

A man says about something he wants to do; 'it cant be wrong because it doesn't do anyone else any harm'…He is thinking it does not matter what his ship is like inside provided that he does not run into the next ship…He quite understands that he must not damage the other ships in the convoy, but he honestly thinks that what he does to his own ship is simply his own business. But does it not make a difference whether his ship is his own property or not? Does it not make a difference whether I am, so to speak, the landlord of my own mind and body, or only a tenant, responsible to the real landlord? If somebody else made me , for his own purposes, then I shall have a lot of duties which I should not have if I simply belonged to myself (p.74).

Quite a profound illustration: We never made ourselves

and we do not comprehend many things about ourselves. We never invented consciousness. We all have it and we

observe it in other people and are aware that it makes our body machinery to work; it very feels like our own property which we can use as we desire, but then we do not even know what this consciousness really is or when and how it came to be. What is more, there was never prior consent for its operation in *us*. We certainly are not the 'patent' holders of this 'intellectual property'; we certainly are not the 'landlords' of *our* bodies or souls! It's a mystery for which the explanation of God being the 'landlord' (owner) is the most reasonable. If this God exists (I firmly believe he does), then it makes profound sense to say that he was the 'manufacturer' who set the 'rules of operation' of the human machinery. Yes, he set the moral compass in human consciousness and is therefore the standard by which things stand to be good or evil, right or wrong.

Materialism versus Morality

Notice that *good* or *evil, right* or *wrong,* and *beautiful* or *ugly* are terms known only by conscious minds; a piece of furniture can never be aware of its looks; it is a mere assembly of mindless atoms. It is the human mind that looks at a particular arrangement of a chunk of matter and describe it as 'beautiful' or 'ugly'. Without human consciousness everything else would be merely mindless processes, movements, collisions and different arrangement of atoms.

So, if, as is believed by materialists, humans are nothing more than a material phenomenon then we would not even know what we are discussing right now because knowledge, contemplation and intentionality are clearly not at the level of matter. Furthermore, if materialism is true then what Adolph Hitler did by torturing and killing Jews was not wrong. Hitler was simply an ensemble of atoms that happened to be that way; he was merely dancing to the tune of his DNA. In such a world the words of the leading atheist Richard Dawkins would be true - *"there is at bottom no design, no purpose, no evil, no good, nothing but pointless indifference. . . We are machines for propagating DNA . . . . It is every living object's sole reason for being"* (Dawkins, 1996). But this obviously doesn't strike with reality. Dawkins himself turns around and calls religion evil. So, how does one person say evil does not exist and then denounce religion to be evil? As was pointed out by Lewis (1947), naturalists will say that all ideas of good and evil are hallucinations – shadows cast on the outer world by the impulses which we have been conditioned to feel – but in a moment later you will find them exhorting us to work for posterity, to educate, revolutionise, liquidate, live and die for the good of the human race. This again reminds me of Francis Crick when he said that *"you, your joys and your sorrows, your memories and ambitions, your sense of personal identity and free will, are in fact no more than the behavior of a vast assembly of nerve cells and their associated molecules"* (1994, p.3). This was quite a strong assertion but which immediately implied that his 37 words you have just read were blips and (I would say) misfirings of neuro-circuits, but which expect other circuits to believe *them* to be conveying truth; if the human mind is no more than a product of neuro-circuits, how come we reason, not only within ourselves but with each other, and also have moral intentions? The plain truth is that there is more than meets the eye to the human assembly of nerve cells and molecules. Yes, there is a dimension to life which is beyond matter and energy. It is a spiritual dimension and that is where morality resides.

Objective Moral values

When terrible inhumane acts are performed even materialists seem to suspend their philosophical gymnastics in denouncing the evil acts. There is always this palpable feeling of standard or acceptable ways of how a human being *ought to* behave. This invisible social standard permeates different cultures and societies. Such moral values which stand true regardless of time and society are said to be objective. Professor Craig (2010) explains:

To say that there are objective moral values is to say that something is good or bad independent of whatever people think about it. Similarly, to say that we have objective moral duties is to say that certain actions are right or wrong for us regardless of what people think about it. So, for example, to say that the Holocaust was objectively wrong is to say that it was wrong even though the Nazis who carried it out thought that it was right, and it would still have been wrong even if the Nazis had won World War II and succeeded in exterminating or brainwashing everybody who disagreed with them so that everyone believed the Holocaust was right.

Clearly there is something deep inside every one of us (including atheists) that resents injustice or atrocity. The question is, if there is no God, by what standard do we judge something to be good or evil? Think of this: would there be such a thing as north or south if no objective compass was agreed on? In a world without a compass all concepts about direction of north, south, east or west collapse as there will be no reference point. Being in the South Pole will all depend on one's pinion. But please note that while a compass is a product of human convention, the same cannot be said of morality. Morality cannot be a mere emergent characteristic of human society; a product of evolution as atheists would say.

The mechanism of evolution by natural selection works to prolong favorable attributes through a selfish game of survival of the fittest. So, some evolutionists have explained that altruism (an aspect of behaviour that doesn't seem to be in agreement with our "selfish genes") is not truly an objective virtue but simply a behaviour which exhibits itself in hope of a favorable return – 'scratch my back and I will scratch yours'. This explanation implies that altruism is also an exhibition of selfishness but masquerading as a virtue of charity. While this explanation is true in a number of situations it has failed to account for genuine acts of love which manifests in people who have selfishly done non-rewarding charitable works, some even at the peril of their own lives. History is replete with such personalities. All this shows that there is a moral compass deeply embedded in the fabric of human consciousness, and the one sensible explanation for that morality is God. Simply put we are moral beings because we were made in the image and likeness of God. And it is by his standard

Immorality – a result of losing identity

It should now be clear why the world is experiencing a general decadence in morality. The root cause of the problem is man forgetting his origins in God. The creator, and hence the *purposer* of life, should have been the compass to direct human life in the world; without that compass, man is a voyager without direction and without a sense of ultimate destination. In essence his existence becomes a meaningless pursuit for *more*, a bottomless hole which never gets filled. In such a Godless world, there is no ultimate purpose nor ultimate justice; even a person who commits suicide (provided he leaves all the money required for his funeral expenses and for the children he leaves behind) does nothing wrong; he owns his life and therefore has the right to end it. Oh! What vanity!

A loss of knowledge about one's origins clearly leads to wrong identity. Wrong identity in turn breeds wrong actions. Think about a lion cab which is born and raised among sheep. It will want to eat, live and behave like sheep. I believe the same thing has happened to people. It all began when one man on his voyage around the Galapagos Islands began to study and compare animal species. He noticed similarities among different living things and then posited that chimpanzees are man's cousins. Indeed many believe that human beings are just another kind of animal species. So, is it not now clear why modern man desires to create a human zoo where fornication is simply copulation, and nudity simply being natural?

Materialism clearly ferments a morally bankrupt society. I could not agree more with Amit Goswami, a University of Oregon physicist, when he said that *"just as global warming is endangering our world, urgent social problems are growing that cannot be solved within the materialist approach; in fact for most of these problems, materialism is the root cause"* (Goswami, 2012, p.53). Surely, if people believe right they will live right and if they believe wrong they will live wrong; we believe and therefore we live. One of these days, sex is going to be taken as a simple social act that there will be people saying there is nothing wrong to copulate with anyone as long as the two have consented. One wonders just what would be the limit to such 'freedom' – as there will be no injury inflicted, would it be alright for a brother to enjoy the pleasure of copulation with his sister, or a father with his daughter and a mother with her son? I once heard a university professor proudly acclaim that there would be nothing wrong with that. What would I be left to say but that education can always make a person intelligent but not wise. Indeed, we have many educated idiots around.

If there is no such a thing as the author of marriage who in the beginning decreed that conjugal relations should only exist within the bounds of holy matrimony, then all these things are permissible. Surely a people who cannot distinguish between a male's chest from a female's will also not distinguish sexual intercourse from a handshake.

7.
Pain and Suffering
Just where is the benevolent creator?

"How can we be sure that God does not exist? Perhaps there's a reason why God permits all the evil in the world. Perhaps it somehow all fits into the grand scheme of things, which we can only dimly discern, if at all. How do we know? As a Christian theist, I'm persuaded that the problem of evil, terrible as it is, does not in the end constitute a disproof of the existence of God. On the contrary, in fact, I think that Christian theism is man's last best hope of solving the problem of evil"

William Lane Craig

December 2014, I was walking across a steep cliff where a typhoon had ravaged one side of a hill destroying cabins and killing people that were living around the area. Some people had run for safety into church buildings but which later filled up with flood waters; the typhoon had not spared the sanctuaries! This was in *Cagayan de Oro* city of the Philippines.

Back home in Zambia, I had once watched news reports of typhoon *Haiyan* which killed over 6000 people. It affected the Philippines, Micronesia, Vietnam and southern China. Watching the news, I was so filled with grief at the sight of helpless little children who were trying to escape the catastrophe. However, being there and always having to be alert to news updates about when and where the next wave will hit was quite an unpleasant experience; I thought about the pain and anguish of families which had lost their loved ones to the floods.

While some people in moments of such despair would sob in prayer and comfort one another of the spiritual hope, others have been led to either question whether God exists, or, if he exists, what would be his purpose in letting such suffering occur. Many have wondered how a benevolent all-knowing and all-powerful creator can be silent amidst all such problems: could a good God keep silent at the sight of a poor little child gasping for life? Why the silence when a faithful, good and humble devout believer is dying with cancer? The famous philosopher and unbeliever, David Hume, probed:

> *Is he willing to prevent evil, but not able? Then he is impotent. Is he able, but not willing? Then he is malevolent...Why is there any misery at all in the world? Not by chance, surely. From some cause then. It is from the intention of the deity? But he is perfectly benevolent. Is it contrary to his intention? But he is almighty* (cited in Plantinga, 1974).

Well, the list of realities which seem incompatible with the existence of an omnipotent creator can be endless. I should say that there are no simple answers to these questions. Even so, atheists give a very simplistic approach to the problem - *"in a universe of electrons and selfish genes, blind physical forces and genetic replication, some people are going to get hurt, other people are going to get lucky, and you won't find any rhyme or reason in it, nor any justice. The universe that we observe has precisely the properties we should expect if there is, at bottom, no design, no purpose, no evil, no good, nothing but pitiless indifference"* (Dawkins, 1995). Theism on the other hand presents an alternative picture of hope. I find this hope not to be a mere crutch for fooling our human feelings of despair (as it has been said by unbelievers) but one which seems logically coherent with the cumulative evidence presented in the past chapters: we saw how everything, from the initial conditions of the universe, through the different stages of its development, was so finely-tuned as if to have an ultimate goal of bringing forth life. However, we now encounter a problem of how this apparent final intention of fine-tuning – *life* – is always threatened by evils of sickness, crime, natural disasters, etc. In the atheistic view, the long journey we have taken through the marvels of how everything seems to be an encapsulation of the wit of a superintellect ends in a disappointing tale of illusion; we are simply in a universe of electrons and selfish genes where there is actually no rhyme, reason, nor any justice. However, the weight of evidence of design and purpose so much surpasses this apparent meaninglessness of evil that it behooves us to ponder and reflect deeply on what could be a more sensible explanation.

In dealing with the problem of pain and suffering it is first important to be aware that there two kinds of evil: the evil caused by the cruelty, arrogance and just plain foolishness of man (hereafter called *moral evil*) and the one caused by nature such as earth-quakes, hurricanes, diseases, etc. (hereafter called *natural evil*). Now, as you will shortly see, the existence of moral evil being possible in the presence of an almighty God can easily be explained using the concept of free will. However, the problem of natural evil poses a challenge. Let me explain.

Moral evil and free will

As human beings we have found ourselves in a world where all other things, except ourselves, are governed by precise laws for which any slight alteration would cause chaos. On earth this has worked very well to the benefit of man. Imagine what kind of a world it would have been if water (just to mention one example) was conscious and had free will? Imagine the drama and horror of drinking water which suddenly - during its movement through your throat - decides to have a 'transgender' feeling of being acid? Imagine two oxygen atoms which are bonded to carbon suddenly going into an argument and then one deciding to part company, leaving *his* family as *carbon monoxide*; before you wonder how ridiculous and petty these particles can be you would have suffocated. In such a world human beings would undoubtedly be unhappy and blame God for the mess. But, would not also all forms of matter now, if they had consciousness, condemn the two-legged upright creatures for wreaking havoc to this planet because of their selfish quest for domination and resources?

In the beginning everything operated according to God's perfect will. Of all God's creatures on earth, human beings were the only ones made in his image and likeness, possessing free will. However, note that good as this was, the granting of free will had an intrinsic implication of making evil possible. But this was inevitable in a world where true love and fellowship was to exist. When man later chose the wrong way in exercising his curiosity, God let him freely abandon the right way; man was free to partake of 'the forbidden tree' but he had to live through the 'fruit' (i.e. consequences) of his choice – "*For that they hated knowledge, and did not choose the fear of the LORD: They would none of my counsel: they despised all my reproof. Therefore shall they eat of the fruit of their own way, and be filled with their own devices*" (Proverbs 1:29-31).

Someone may ask: "But why should God deliberately let man indulge in wrong?" A *vain* exercise it would seem, but not exactly so. The presence of free will where the creator promises future redemption for *wrong* (sin) can be rationalized if there is an intention to let the *vanity* expose the foolishness of the wrong doer, and let this process be his learning experience. It can thus be explained that God was willing to lose creation to *vanity* in hope that sentient beings will through experience come to freely choose God's perfect will and through this get redeemed:

> *For the creature was made subject to vanity, not willingly, but by reason of him who hath subjected the same in hope, Because the creature itself also shall be delivered from the bondage of corruption into the glorious liberty of the children of God. For we know that the whole creation groaneth and travaileth in pain together until now. And not only they, but ourselves also, which have the firstfruits of the Spirit, even we ourselves groan within ourselves, waiting for the adoption, to wit, the redemption of our body* (Romans 8:20-23).

Letting man experience the vanity and problems that would result from his short-sighted (myopic) will, appears to be the one way God could present an opportunity for people to do his good will but yet without imposing it on them.

Someone may still resent – "How could an all-powerful God not create creatures who would automatically do right things and by this prevent the problems of evil that would result from sin? Doesn't the holy book say all things happen according to his will?"

Well, two questions to consider in order to handle this difficulty:

> *First*, is it possible for evil to happen because of God's will and at the same time the same evil being contrary to his desire?

> *Second*, is it possible to have a creature of free will but which is automated to do right things?

Lewis (1944) gives an illustration of how someone's will can permit something in one way, and yet *it* being not his will in another way;

> *...anyone who has been in authority knows how a thing can be in accordance with your will in one way and not in another. It may be quite sensible for a mother to say to the children, 'I'm not going to go and make you tidy the schoolroom every night. You've got to learn to keep it tidy on your own.' Then she goes up one night and finds the Teddy bear and the ink and the French Grammar all lying in the grate. That is against her will. She would prefer the children to be tidy. But on the other hand, it is her will which has left the children free to be untidy....You make a thing voluntary and then half the people do not do it. That is not what you willed, but your will has made it possible* (p.48).

The important thing to understand in all this is how the law of free will works. Free will makes it impossible for a person to automatically do good. As long as there is free will, then there will be a possibility of the presence of evil just as there will be of good:

> *Some people think they can imagine a creature which was free but had no possibility of going wrong; I cannot. If a thing is free to be good it is also free to be bad. And free will is what has made evil possible. Why, then, did God give them free will? Because free will, though it makes evil possible, is also the only thing that makes possible any love or goodness or joy worth having. A world of automata – of creatures that worked like machines – would hardly be worth creating* (p.48).

Lewis then adds provocative but true words:

> *If God thinks this state of war in the universe a price worth paying for free will – that is, for making a live world in which creatures can do real good or harm and something of real importance can happen, instead of a toy world which only moves when He pulls the strings – then we may take it it is worth paying* (p.48).

Natural Evil

Now, while the argument of free will presents a reasonable argument for the cause of evil in the world, it does not account for evils which arise from non-human activities such as natural disasters. What is more, natural disasters can claim both the lives of believers as well as unbelievers; the lives of infants as well as adults. So, if there is an all-knowing and all-powerful benevolent God, the question is often asked, why would he permit such a situation?

While it is not easy to give a definite convincing answer, I should say that, if all issues I have discussed this far (including the recent research results that demonstrated that life continues to exist outside a human body after death) truly amount to evidence of the existence of God (which I firmly believe they do) then it goes without saying that there is truly a spiritual dimension beyond what our human senses can detect. And if the spiritual world exists, then the Biblical records of the existence of supernatural beings called angels, is also trustworthy. Like human beings these creatures also have free will. And again like humans there are some angels who, under the leadership of a strong principality called Lucifer or Satan, rebelled against God and work as agents of evil. In the beginning when man willfully decided to give heed to Satan, he, so to speak, gave preference to him as the new master and thus the evil principality became *"the god of this world"*. God willingly *"delivered"* the world to Satan at man's choice, albeit for his lesson. However, it has been a bitter experience not only for sentient beings but *"all creation"*. Romans 8:20-23 in the Amplified Bible version seems to be very congruent with this explanation:

> *For the creation (nature) was subjected to frailty – to futility, condemned to frustration – not because of some intentional fault on its part, but by the will of Him Who so subjected it. [Yet] with the hope…We know that the whole creation (of irrational creatures) has been moaning together in the pains of labor until now.*

Notice how the record of the words of Satan in his conversation with Christ to try to tempt him affirms the fact that the world had been delivered to him:

> *And the devil, taking him up into an high mountain, showed unto him* **all the kingdoms of the world** *in a moment of time…And the devil said unto him, All this power will I give thee, and the glory of them: for that* **is delivered unto me***; and to whomsoever I will I give it* [emphasis mine].[16]

---

[16] The New Testament is a record of how Christ came to 'pay' for the redemption of man and the restoration of the 'title deed' to the possession of the world. Although the price has been paid, there is still a span of time during which sons and daughters of God are being 'born' into the kingdom which will one day *come with power* (cf. Ephesians 1: 14). In that day creation *"shall be delivered from the*

These premises make what Saint Augustine said regarding the problem of natural evil make sense. His explanation follows that while moral evil is caused by the free will actions of human beings, natural evil is a consequence of the free will of non-human spirits. If this is true, Plantinga (1974) suggests that *"then natural evil significantly resembles moral evil in that…it is the result of the activity of significantly free persons…both moral and natural evil would then be special cases of what we might call broadly moral evil - evil resulting from the free actions of free beings, whether human or not"*(p.58-59).

Please note that some believers carelessly attribute any occurrence of a natural disaster to God's act of judgment. While there are incidents in the Bible of the antediluvian flood or droughts which had been foretold by prophets as judgments, it is important to be also aware of the Lord Jesus' words concerning some two incidents for which he dismissed assertions of thinking that the victims were being proportioned their sinful wages:

> *There were present at that season some that told him of the Galilaeans, whose blood Pilate had mingled with their sacrifices. And Jesus answering said unto them, Suppose ye that these Galilaeans were sinners above all the Galilaeans, because they suffered such things? I tell you, Nay: but, except ye repent, ye shall all likewise perish. Or those eighteen, upon whom the tower in Siloam fell, and slew them, think ye that they were sinners above all men that dwelt in Jerusalem? I tell you, Nay: but, except ye repent, ye shall all likewise perish* (Luke 13:1-5).

---

bondage of corruption into the glorious liberty of the children of God" (Romans 8:21).

Indeed in the book of Acts we read about how Paul once suffered a natural disaster of a boisterous wind which almost took the lives of people who were aboard with him on a ship. In one epistle he listed a number of his ordeals – both which were caused by people and those by nature:

> *Thrice was I beaten with rods, once was I stoned, thrice I suffered shipwreck, a night and a day I have been in the deep; In journeyings often, in perils of waters, in perils of robbers, in perils by mine own countrymen, in perils by the heathen, in perils in the city, in perils in the wilderness, in perils in the sea, in perils among false brethren; In weariness and painfulness, in watchings often, in hunger and thirst, in fastings often, in cold and nakedness* (2 Corinthians 11:25-27).

In all his writings, Saint Paul expresses the belief that the present world is in the hands of Satan and his horde of demons which work through corrupt mankind (Ephesians 2:2). Him and other apostles proclaimed a hope for the time when Christ shall come and establish his kingdom on earth; a time when sons and daughters of God shall take over the rulership of this world (Revelation 20:1-6, 2 Peter 3:12, Romans 8:21, Daniel 7:27).

Please note that the explanation concerning evil spiritual powers influencing nature is not scientific. From the standpoint of science, it can be said to be an opinion, suggestion or speculation. However, it is also important to be aware of the inherent limit of science in dealing with problems that are not in the realm of physical matter. Science is the way to the truth about the natural world. But for things which constitute the dimension which is beyond space and time, no amount of sophistication of science can explain anything. In a similar manner science is forever incapacitated to model the state of 'nothing' before the existence of matter, space, energy and time. What is more puzzling is that this region of 'nothing' is not really no more; it is still present and active. Notice: if the universe (which includes space) is expanding, common sense should let us know that it is expanding *into* something *unoccupied*. So then, what is this *unoccupied* region? If it is not matter, space or energy, then it is by definition non-physical, and if non-physical then only one word in our limited language vocabulary remains to explain it – supernatural! Here now rises the greatest difficulty of humans trying to fathom the supernatural using natural means. This is by default impossible for "*the things of the Spirit of God…are spiritually discerned*" (1 Cor. 2:13).

So then the challenge; how shall mortals fathom the supernatural? How shall they search for and discover God?

8.
In Search of God
Have you the right attitude?

"The fact which is in one respect the most obvious and primary fact, and through which alone you have access to all the other facts, may be precisely the one that is most easily forgotten – forgotten not because it is so remote or abstruse but because it is so near and so obvious"

Clive S. Lewis

Granted if every atom or living cell tissue had words inscribed "*Made by God*" would it have made everyone a believer? Or if one day we were all to wake up to a big surprise of an inscription in the sky "*I am God, I Exist*", would it convert any unbeliever?

If everyday simple evidence will not convince a person, more or greater of it will never accomplish anything. If today God suddenly manifested himself in a spectacular display for everyone to see, atheists may respond in one of the following ways:

- *Here comes the first proof of extra-terrestrials!*
- *NASA is playing a trick on us.*
- *It's mysterious but a natural phenomenon which science will eventually explain.*
- *Someone has monkeyed with our sense of sight; it's an illusion!*
- *We are a digital simulation, and this could be a program bug.*

The wonderful gift of free will is what makes human beings such interesting creatures. Because of free will many people want to believe what they want. For example many atheists will say that the earth is just too little - a very small infinitesimal speck of dust – for there to be a God who is concerned about it. This argument weighs the importance and hence relevance of something on the basis of relative size.[17] With a simple illustration Lewis (1947) punctured this argument:

> *If from the vastness of the universe and the smallness of Earth we diagnose that Christianity is false we ought to have a clear idea of the sort of universe we should have expected if it were true. But have we?...If it were empty, if it contained nothing but our own Sun, then that vast vacancy would certainly be used as an argument against the very existence of God. Why, it would be asked, should He create one speck and leave all the rest of space to nonentity? If on the other hand, we find (as we actually do) countless bodies floating in space, they must be either habitable or uninhabitable. Now the odd thing is that both alternatives are equally used as objections to Christianity...The doctor here can diagnose poison without looking at the corpse for he has a theory of poison which he will maintain whatever the state of the organs turns out to be* (p.80).

The words of Lewis tell us that many sceptics will always come up with excuses even when a fact is plain. At one time religious people used to be ridiculed at when they would speak about the beginning of time (atheists then believed the universe was eternal). Then came the evidence of the Big Bang validating the Biblical "*In the beginning*". What happened next? A pursuit for models to compete against the Big Bang model surged. Well, for many years now, all of them have failed to challenge the standard model. And what about the recent scientific study which

---

[17] Quite a conspicuous flaw in the argument – imagine if the method was used to ascertain the different levels of importance of body organs; a leg will be more important than your heart or brain!

demonstrated that life continues after death? Notice that all the while atheist scientists have had an idea that all matter including human beings and their consciousness are reducible to atoms and effects produced by interactions and movements of these atoms. In this worldview there is no such a thing as a spirit in man that would exist independently of his physical being. But we now wait to hear theories that will come up following the results of the largest-ever study about life-after-death experiences that were released last year (2014), making headlines on international media (see **Appendix II**).

In 1955, William Branham, a controversial figure and paradox in evangelical Christianity, claimed to always see a halo following him. When this light would overshadow a person he would tell a person's name, disease, and even thoughts. He began what came to be known as faith-healing campaigns in USA which attracted thousands and his meetings were the largest around the world at the time. Of course it's hard to believe such facts in a modern setting where there is so much falsehood. However, several people including unbelievers got converted at seeing the supernatural light. Again this is not convincing enough to someone who wants 'hard data'. Such a one, a critic, attended one meeting where he decided to photograph Reverend Branham when he claimed that the halo had come over him. A camera (a scientific instrument with no emotional or doctrinal bias) was aimed and the shot taken. What came out astounded the world. The image was submitted for scientific investigation. George J. Lacy, president of the American Society of Questioned Documents (an FBI agency), who was also a critic, had this to say after examining the photograph:

You are going to pass out of this world like all other mortals, but as long as there is a Christian civilisation, your picture shall remain. To my knowledge, this is the first time in all the world's history that a supernatural being has been photographed and scientifically validated" (cited in Jorgensen, 2008).[18]

Fig.8.1 *Scientifically validated photo of halo on Rev. William Branham*

To date Rev. William Branham's photo of the halo is held in the Library of Congress.

Now, will this photo convert a critic? Nine times out of ten, No. Proof (in whatever form) will never convince a mind which is determined not to believe what it doesn't want to believe. Many times it is not a quest for evidence but for opportunities to falsify. As Collins, an agnostic turned believer, once testified concerning his life, it was not about the lack of evidence but not wanting the evidence.

Let us take note of two things concerning the matter of believing the evidence of God:

---

[18] See **Appendix III** for George Lacey's report of the investigation on the photo

First – *it is a matter of attitude*

Why would God reveal himself to some people and not others? The basic problem in many who call for the need to prove God has to do with attempting to examine God with wrong sets of tools – test-tubes, telescopes, and microscopes, as though God were matter (the *gods* must be laughing at us!).

C.S. Lewis in *Mere Christianity* excellently put it this way:

> *If you are a geologist studying rocks, you have to go and find the rocks. They will not come to you, and if you go to them they cannot run away. The initiative lies all on your side. They cannot help or hinder. But suppose you are a Zoologist and want to take photos of wild animals in their native haunts. That is a bit different from studying rocks. The wild animals will not come to you: but they can run away from you…Now a stage higher; suppose you want to get to know a human person. If he is determined not to let you, you will not get to know him. You have to win his confidence. In this case the initiative is equally divided…When you come to knowing God, the initiative lies on His side. If He does not show Himself, nothing you can do will enable you to find Him. And, in fact, He shows much more of Himself to some people than to others – not because He has favourites, but because it is impossible for Him to show Himself to a man whose whole mind and character are in the wrong directions* (Lewis, 1944, p.164).

Wise words; finding God or understanding his ways is a matter of attitude. God is not inert matter which you can study, verify and validate in any style of your wish. It is here that the wise of the world often exhibit foolishness, and it is no wonder that God would reveal himself many times to simple minds – "*For ye see your calling, brethren, how that not many wise men after the flesh, not many mighty, not many noble, are called: But God hath chosen the foolish things of the world to confound the wise; and God hath chosen the weak things of the world to confound the things which are mighty*" (1 Cor.1:26-27). Surely God does reveal himself to those who are sincere and humble enough. It is at this point that talk about the supernatural becomes a matter of personal experience which cannot be duplicated, imitated or even scientifically validated.

Second – *it is a matter of attention*

Is it truly the case that God is silent about his existence, or is there not already too much noise in our environments crying out in giving evidence, in fact, so loud that our senses have become too numb to realise that everything is actually a great miracle? Here again Lewis (1947) gave a useful illustration on how our senses sometimes tend to become unconscious to the most obvious things:

> *To some people the great trouble about any argument for the Supernatural is simply the fact that argument should be needed at all. If so stupendous a thing exists, ought it not to be obvious as the sun in the sky?...We must notice two things: When you are looking at a garden from a room upstairs it is obvious (once you think about it) that you are looking through a window. But if it is the garden that interest you, you may look at it for a long time without thinking of the window. When you are reading a book it is obvious (once you attend to it) that you are using your eyes: but unless your eyes begin to hurt you, or the book is a text book on optics, you may read all evening without*

once thinking of eyes...these instances show that the fact which is in one respect the most obvious and primary fact, and through which alone you have access to all the other facts, may be precisely the one that is most easily forgotten – forgotten not because it is so remote or abstruse but because it is so near and so obvious...The Supernatural is not remote and abstruse: it is a matter of daily and hourly experience, as intimate as breathing: Denial of it depends on a certain absentmindedness (p.65).

I believe that an alert mind – free from the noise of prejudice and self-conceit - will be able to discern the presence of God in the very nature that we see around us in everyday life – *"Be still, and know that I am God"*, reads the ancient psalm (Psalm 46:10).

Hearing the small still voice

If a person's mind is confined within a straightjacket of only believing things which can be proved by science, he or she will surely be ignorant of many improvable things such as love and the purpose of human life. *"The existence of a limit to science"*, says Sir Peter Medawar, *"is...made clear by its inability to answer childlike elementary questions having to do with first and last things – questions such as 'How did everything begin?'; 'What are we all here for?'; 'What is the point of living?'"* (Medawar, 1979, p.31). Such things are inherently unreachable by tools of empirical proof. However such knowledge can be accessed by *revelation*. What is revelation?

Consider this: Someone finds a written note lying on the ground. The note is in an unknown language. Taking it to the laboratory, a chemist can perform all types of conceivable examinations of the paper or type of ink which was used for the writing; all such methods and tools will never come anywhere close to decoding the meaning of the message. It is at this point that science becomes lame and *revelation* matters. The author of the letter will have to explain (*reveal*) his mind about the purpose and meaning of the writing. The lesson here is that, just as you do not get to understand the meaning of a message written on paper by examining the chemistry of the ink used or the medium on which it was conveyed, so you can never understand the meaning of life or the universe by examining its physics or chemistry of atoms and molecules. Likewise unbelievers often use wrong sets of tools in probing the existence of God. It is Albert Einstein who once said, "*I want to know the mind of that One*", meaning God. On he went with the tools of physics and mathematics to decipher mysteries of this physical world. He remarked:

> *The human mind is not capable of grasping the Universe. We are like a little child entering a huge library. The walls are covered to the ceilings with books in many different tongues. The child knows that someone must have written these books. It does not know who or how. It does not understand the languages in which they are written. But the child notes a definite plan in the arrangement of the books - a mysterious order which it does not comprehend, but only dimly suspects.*

And in another place he said:

> *Everyone who is seriously involved in the pursuit of science becomes convinced that a spirit is manifest in the laws of the Universe - a spirit vastly superior to that of man, and one in the face of which we with our modest powers must feel humble.*

It is interesting to note that whenever mortals behold the wonder of the universe and nature, and the precise laws that govern them, there is always that religious feeling of awe and reverence. In the silence of meditation over the power that upholds all things, one feels small and the large earth shrinks to pettiness. In December 1968 the world held its breath as people around the world waited to hear news about three men who were on the first manned spacecraft to orbit the moon. Frank Borman, James Lovell, and William Anders took the first photos that showed the rising of the earth over the moon. The feeling of being in space, away from the now small and fragile-looking earth, must have been very awesome to the three astronauts as they did the unexpected: they jointly read the first ten verses of the book of Genesis and this was broadcast through a live television transmission. This event got the attention of one unbeliever – Francis Collins. He recounts:

> *As an agnostic on the way to becoming an atheist at the time, I still remember the surprising sense of awe that settled over me as those unforgettable words – 'In the beginning, God created the heavens and the earth' – reached my ears from 240,000 miles away, spoken by men who were scientists and engineers, but for whom these words had obvious powerful meaning* (p.160).

The sight and phenomenon of, not only celestial bodies in space but, everything in nature around us seems to speak something. The voice is still but when a person hears it, it can change everything about himself and ignite the sense of hope and ultimate purpose of life. For different people it has taken different things to awaken them to this reality; for some it's suffering and others an imminent danger.

In the Judeo Hebrew scriptures we read about an ancient noble man who suffered a deep anguish of soul when misfortune fell on him. In a moment he lost his possessions, his children and later his health. At first he encouraged himself to endure, believing that the God who gave him the wealth had decided to take away the things. But as days turned into weeks and weeks into months, the petrifying sores and boils over his body became unbearable. He began to reason and question the justice of God in letting him experience all the misfortunes. He dared God to point out his unrighteousness. One day an odd thing occurred to him; an unusual whirlwind approached and out of it came a resounding voice that shook the loins of his mind. In a moment he felt his unrighteousness, unworthiness and foolishness. He trembled and confessed:

> *I uttered that I understood not; things too wonderful for me, which I knew not…I have heard of thee by the hearing of the ear: but now mine eye seeth thee. Wherefore I abhor myself, and repent in dust and ashes*" (Job 42:3-6).

As human beings we are veiled in flesh and our reality is so very limited to the senses. A simple glimpse into the supernatural shakes loose what we thought we had knowledge of: Job had been complaining for quite some time but it took that short moment of the manifestation of the supernatural for him to abhor the words he had been speaking. "*I understood not*", he spoke in trembling.

God manifested in a whirlwind to Job but the whirlwind was not God. Another man with a similar experience was Elijah. The story is recorded in the book of 1 Kings chapter 19:9,11-13.

And he came thither unto a cave, and lodged there; and, behold, the word of the LORD came to him, and he said unto him, What doest thou here, Elijah?... And he said, Go forth, and stand upon the mount before the LORD. And, behold, the LORD passed by, and a great and strong wind rent the mountains, and brake in pieces the rocks before the LORD; but the LORD was not in the wind: and after the wind an earthquake; but the LORD was not in the earthquake: And after the earthquake a fire; but the LORD was not in the fire: and after the fire a still small voice. And it was so, when Elijah heard it that he wrapped his face in his mantle, and went out, and stood in the entering in of the cave. And, behold, there came a voice unto him, and said, What doest thou here, Elijah?

Granted if there were some curious rational unbelievers, they would have gone and investigated the claim of the Hebrew prophet by examining the air, wind or ashes of fire to verify the presence of God. Such are the ways of the foolishness of man; trying to investigate God or the purpose of the world and life by examining its chemistry and physics. It should be a simple fact that when you meet a person and would like to know him, examining his outfit or looks may never bring you anywhere close to knowing who he really is. That can only happen if you win his confidence for him to speak. Likewise, to the sincere, humble and such as would learn, the small still voice of God can speak just as it did to Elijah. When that voice spoke, it asked him, *"What doest thou here, Elijah?"* The revelation of the voice of God to a person is the beginning of understanding the purpose and meaning of why he is here on earth. The important question that each person needs to ask is, "Do I have the right attitude for God to reveal himself in my life or am I being too wise in my own conceit?" True and meaningful knowledge of God can only come through revelation.

Be admonished that "*the fear of the LORD is the beginning of knowledge, but fools despise wisdom*" (Proverbs 1:7). What are fools but such as would look into a mirror, behold their existence but yet deny the existence of their maker - "*The fool hath said in his heart, 'There is no God'*" (Psalm 14:1).

Questions and Answers

The following are some questions and statements atheists have sent through email correspondence and posts on my blog.[19] Some words have been edited, re-phrased or combined with other similar ones for purposes of clarity. The answers I present here are not fully exhaustive but only convey core points.

1. *"If you were born in Afghanistan, you would be a Taliban and would have bombed for your God."*

Well, let us try extending this statement by asserting that, if I were an acquaintance of Charles Darwin and was with him on his voyage around the Galapagos Islands, I would have become a Darwinist and therefore the theory of evolution is false because I merely believed it through my association with its proponent. Absurd logic isn't it? How can the truthfulness or falsehood of a fact be determined by *where* someone has grown up or lived to believe it? While it is true that an environment can influence a person's belief, assessment of the truthfulness of the belief should be based on the evaluation of the facts claimed by the belief system and not the examination of influences that may have led to the belief. For example Michael Faraday a British physicist and chemist was a deeply religious man; his obsession of science was influenced by

---

[19] www.andrewcphiri.com. Note that there is slight editing on spellings and grammatical errors of some questions, and the answers presented here have a little more details than what you may see on the website.

his belief in God as the underlying power that unified everything in nature. His faith inspired him to perform a lot of scientific experiments as methods for deciphering God's handiwork in nature. While someone may disagree with Faraday's religious convictions it is worthwhile to know that they are what led him to discover *electromagnetic induction* and *electrolysis*. Now, would it be reasonable to discard electromagnetic induction and electrolysis on the basis of Faraday's religious influences that led to their discovery?

It is true that science and faith do not use the same instruments of empirical verification, however, it is important to see that they both present facts which can be assessed by the guideline of following where the evidence leads. So, although someone may have grown up in a region where a certain religion is practised, as he grows up and gets exposed to knowledge, he should make an informed decision concerning his faith. If a person calls himself a believer but without any reason for his beliefs, apart from carrying on what his family, race or people are accustomed to, then he or she does not truly believe. His faith is blind and a product of adaptation. Although there are many people who are Christians in this way, it is important to be aware that there are also some whose faith is based on evidence. That is why there are some Christians who have afterward converted to Islam and others who grew up as well-trained Islamists later converted to Christianity. Likewise, there have been some atheists who later converted to theism (Clive S. Lewis, Anthony Flew, Francis Collins, just to mention three). In essence that is what makes human life interesting – we grow up and later decide for ourselves what we think is the best explanation of things around us.[20]

2. *"We still do not know how life started but we know how it progresses. We are working on this one, but we won't substitute our not knowing for something delusional."*

Who are the *"We"* in this statement? Is it all scientists who take faith in God as a delusion? Does the *"We"* include the famous neurosurgeon, Ben Carson? Does it include the famous geneticist and leader of the Human Genome Project, Francis Collins? Does it include the Nobel laureate astronomer Arno Penzias? – Only to mention a few famous scientists who believe that available evidence leads to the conclusion that there is God. This is their view. There are also other equally intelligent scientists like Richard Dawkins, Sam Harris and Daniel Dennet but who subscribe to atheism. This situation simply shows that the real conflict is not between science and atheism but between different worldviews, otherwise all men of science would have been atheists. Science neither approves nor disapproves of religion; it is mute. Science will only inform you that water is made up of two atoms of hydrogen bonded to oxygen. One scientist can pick it from there and say that the initial cause of fundamental particles that led to the existence of these atoms was God, and another scientist may believe that they just began to exist on their own without a divine cause. What is at conflict

---

[20] I am aware that there are some sections of religion which are intolerant and would force and indoctrinate a society into a religion. That is a bad thing and I trust you are aware that not everyone subscribes to such methods.

here is not science but the interpretations of facts established by science. Unfortunately some people will hear a view or statement expressed by a scientist and then believe it to be a scientific statement. Lennox (2009) explains it nicely:

> *The fact that there are scientists who appear to be at war with God is not quite the same thing as science itself being at war with God. For example, some musicians are militant atheists. But does that mean music itself is at war with God? Hardly. The point here may be expressed as follows: Statements by scientists are not necessarily statements of science* (p.19).

So, for example, one should be able to distinguish scientific facts presented by Stephen Hawking from his personal atheistic views. Likewise, Isaac Newton was a deeply devout theologian but this should not make any atheist throw away his laws of motion. If you are a scientist who is also an atheist and therefore regard all believing scientists as non-scientists, you are not different from a man who wears blue sun-glasses and then supposes the whole world to be blue.

A statement of a scientist could be something arising from what he believes about what science has declared: a fossil of a bone structure which resembles a human being and also has some features of a chimpanzee can be unearthed; one scientist will take the similarities to mean that humans evolved from apes, but another will dismiss such a view as plain nonsense; he may believe that humans never evolved; they were created by God as they were. Notice that both these views cannot be proved. The evolutionist has never witnessed any chimp turning into a monkey just like the creationist was never there to witness God's acts of creation. What this goes to show is that both atheistic and theistic views are based on faith. This faith feeds on evidence. The evolutionist's evidence, in this case, are the similarities of bone structures, and for the creationist, his faith rests on the fact that the information code in the DNA's of different creatures show that intelligence put it there as all known forms of information originate from intelligence. So then, how do we decide who is right? The reasonable thing to do would be to go for the evidence that makes sense.

3. *"Adam and Eve were on earth 6000 years ago, according to chronological data presented in the Bible, but archaeology has revealed fossils of dinosaurs and apes which date back to millions of years ago. Moreover Cosmology has convincingly demonstrated that the universe is about 13.7 billion years old. So, hasn't science made the Bible obsolete?"*

It is true that fossil evidence testify that apes, dinosaurs and other prehistoric creatures once existed on earth thousands of years before the time of Adam and Eve. My affirmation of this may sound strange to some Christians who have never taken serious consideration of the facts. But what many do not know is that the Bible actually consistently agrees with paleontological and cosmological data, contrary to the assumption of the question above.

I am aware that many Christians believe that Adam and Eve and their animals were the first forms of life to live on earth. Some would even unwisely say "I would rather believe what the Bible says than what fossils reveal."

Knowledgeable Christians who are aware about the reliability of scientific methods used to ascertain the age of fossils[21] find this attitude quite embarrassing. It is important to know that what nature reveals can never contradict God's Word. Like Galileo said, *"two truths can never contradict each other"*; this renowned Italian scientist advised that wise interpreters of the Bible should strive to find the true meaning of scriptural passages agreeing with those physical conclusions[22]. So let us see how scientific data reconciles with scriptures.

Skull of Pachycephalosaurus at the Oxford University Museum of Natural History. Credit: Wikimedia

Artistic illustration of a dinosaur
Credit: Don Stewart

---

[21] See **Appendix IV**
[22] ofcourse this is pabulum to those who have not yet come to terms with the divine inspiration of scriptures

The Hebrew apostle John, uneducated and without knowledge of the Big Bang, Inflationary and BGV theories, wrote that *"In the beginning was the word, and the word was with God, and the word was God."* The question is: what was this beginning? Was it 6000 years ago when the first parents of mankind appeared on earth? Not so. Genesis 1:28 (in the King James Version) gives an important clue: in this verse God tells Adam and Eve to *"multiply and replenish the earth"*. To replenish is to *re-populate*. This word immediately raises the question of how could first creatures on earth have been commanded to re-fill it. The small word – *replenish* - informs the reader that the earth was once occupied, then emptied and Adam was now required to re-fill it. Notice, to say Adam and Eve were on earth some 6000 years ago is correct, but to say the same amount of years apply to the age of the earth, the universe, apes, dinosaurs and other prehistoric creatures is plain wrong. Leakey (1990) a specialist in African prehistory wrote:

> *In Africa there have been several important discoveries that illustrate the presence of Homo sapiens on the continent for more than 100,000 years; Rhodesian man, discovered in 1920 at Broken Hill[23], in what was then Northern Rhodesia[24], has been regarded as the oldest Homo sapiens found in Africa and has been dated to about 35,000 years ago...The best-known pre-sapiens form of Homo is that which has been attributed to a wide ranging, diverse morphospecies known as Homo erectus. This species was first identified in the Far East and China, but the same form has been collected in North and East Africa...these*

---

[23] Broken Hill, now known as *Kabwe*, is a small town in Zambia's Central Province.

[24] Northern Rhodesia was the name of Zambia before it gained independence from British Colonial rule in 1964.

> *fossils have been dated as being between 1,500,000 and 500,000 years old* (p.194).

While some paleontologists invalidate the inspiration of scriptures on the basis of these findings, some Christians wrongly dismiss the validity of the facts. And it is sad that the Bible has for long been dismissed and attacked by skeptics on the premise of traditional interpretations imposed on it. Unfortunately, when a distortion or lie is repeated over and over again it tends to become fact for many people. However, there are good reasons to believe that biblical data reconciles with paleontological facts of a world that existed before Adam's era.

Now, give close attention to the way the Genesis account of creation begins: "*In the beginning God created the heaven and earth and the earth was without form, and void and darkness was upon the face of the deep*" (Genesis 1:1). These words describe the earth in a disordered state - without form, void and in darkness. The question is, what brought about this dismal condition? Certainly, God did not create chaos and later decide to put it in order. A prophecy of Jeremiah mirror the opening words of the account in Genesis and provides useful details which fit well in the compressed account of Moses. Jeremiah 4:

> 23 *I beheld the earth, and, lo, it was without form. and void; and the heavens, they had no light.*
> 24 *I beheld the mountains, and, lo, they trembled, and all the hills moved lightly*
> 25 *I beheld, and, lo, there was no man, and all the birds of the heavens were fled.*
> 26 *I beheld, and, lo, the fruitful place was a wilderness, and all the cities thereof were broken down at the presence of the LORD, and by his fierce anger.*
> 27 *For thus hath the LORD said, the whole land shall be desolate; yet will I not make a full end.*

These verses are replete with 'spiritual fossils' which have gone unobserved by theologians and were only brought to light by a humble pastor from Jeffersonville Indiana, the late Raymond Jackson. In his 1990 *Contender* periodical, titled *The Testing and Fall of Satan*, Jackson made a very profound remark (but unpopular in mainstream Christianity) when he said *"God did not create this planet in a void, desolate, formless state of total darkness. It was like that because of something that had taken place long before what the account beginning in Genesis 1:2 relates"*(p.2)[25].

The above verses from the book of Jeremiah bring to light what caused the dismal condition on earth: In verse 23 the dismal condition is described using the same words found in Genesis 1:1 - *without form*, *void* and *filled with darkness*. Verse 24 goes further to show what brought about the condition - the quaking of the earth and mountains. Overwhelming paleontological evidence show that for close to two hundred million years dinosaurs had roamed the earth until their sudden mass extinction about 60 million years ago. Joseph Shklovsky, a Russian astronomer, in 1956, postulated a theory which attributed the sudden demise to a single catastrophic event caused by a supernova. However, the scientific community could not take his hypothesis seriously as it lacked convincing evidence. It was later in 1981 when physicist Luis Alvarez, with his son geologist Walter Alvarez, published a similar theory with impressive evidence supporting the hypothesis. The duo discovered a layer of clay with a lot of iridium. This element is very rich in space and rare on earth. These findings led the Alvarezs come to a remarkable conclusion of attributing the iridium deposits

---

[25] Being not clad in the modern glamorous style of Christianity, it is not surprising why Rev. Jackson`s profound teaching on the subject did not attract public attention.

to a meteorite collision with the earth. The duo's report (Alvarez et al., 1980) stated:

> *Our hypothesis suggest that an asteroid struck the earth, formed an impact crater, and some of the dust-sized material ejected from the crater reached the stratosphere and was spread around the globe. This dust effectively prevented sunlight from reaching the surface for a period of several years, until the dust settled on earth. Loss of sunlight suppressed photosynthesis, and as a result most food chains collapsed and the extinctions resulted* (p.1105).

The Alvarez hypothesis was re-affirmed in 1991 by the discovery of a 180km diameter structure buried on the Yucatan Peninsula of Mexico. This crater "presumably produced the largest impact crater on Earth [*and*] must have caused a mass extinction". The graphic description of this mass extinction in an article *Why Did the Dinosaurs Die Out* by the History Channel (2013) is fascinating:

> *Scientists believe the bolide…struck the earth at 40,000 miles per hour and released 2 million times more energy than the most powerful nuclear bomb ever detonated. The heat would have broiled the earth's surface, ignited wildfires worldwide and plunged the planet into darkness as debris clouded the atmosphere. Miles-high tsunamis would have washed over the continents, drowning many forms of life. Shock waves would have triggered earthquakes and volcanic eruptions.*

These facts seem to be perfectly congruent with the record of Jeremiah 4:23-27. Verses 25 and 27 bring out two more informative facts: *first*; the prehistoric catastrophe did not *just* occur; something happened that aroused the anger of God and caused him to judge the earth, and *second*; even though God judged and brought an end to the prehistoric age, he promised not to make a full end. And, true to his Word, God later restored the earth, during which time man was created to govern it.

4. "Give me the evidence that a designer exists and I will solemnly change my views."

This book has been about giving evidence. And in chapter eight I stated that if simple everyday evidence will not convince a person, more or greater of it will accomplish nothing. If today all atoms and specks of dust suddenly manifested the words 'made by God' an atheist could respond in one of the following ways:

- *Here comes the first proof of extra-terrestrials!*
- *NASA is playing a trick on us.*
- *It's mysterious but a natural phenomenon which science will one day explain.*
- *Someone has monkeyed with our sense of sight; it's an illusion!*
- *We are a digital simulation, and this could be a program bug.*

The wonderful gift of *free will* is what makes human beings such interesting creatures. Because of the free will many people want to believe what they want to be right.

Now, with reference to this question, let's look at one evidence that has made many great scientists solemnly believe that while we may not understand everything about God, he is the best explanation of why we and everything else exist:

A series of codes which encapsulate a semiotic

dimension which conspicuously intends to communicate or inform a process has a definite intelligent source – as far as all our human experiences have demonstrated. It is for this simple reason that the SETI project which NASA has heavily funded (despite no results to-date!) hopes to find some signal which will indicate extra-terrestrial intelligence. In the words of Carl Sagan, even a single signal will be enough for us to know that we are not alone in the universe. Well, where is logic in this endeavour – going to outer space to search for signals which would denote an intelligent source when right here on planet earth we have DNA, a molecule that contains precise information that guides the development of organisms? Information IN ALL ITS KNOWN FORMS never arises on its own. ALL EVIDENCE in this world shows that the only source of information is intelligence. The question is, who or what encoded the DNA? And the greater question is, who informed the beginning of DNA itself? If life came from inert matter, how did the first DNA information pop up into existence? Moreover, what caused the first atom to come forth out of nothing? When my answer is God, I am not giving a "God of the gaps" explanation. I am actually applying reason to fathom that if all information in my world has always originated from intelligent minds, then it would be logical to extrapolate the idea and infer that in the beginning was information that informed the Big Bang. Please take note that one may not fully comprehend the nature of this *being* called God but can still accept his possibility just as all physicists believe energy exists - they can measure, assess and quantify it - but yet none can really explain what it is. It is in the light of the foregoing that Ben Carson said "*I don't have enough faith to believe that something as complex as our ability to rationalize, to think and fend and have a moral sense of what is right and wrong just appeared"*. All this shows that it takes just as much faith to believe that there is no God just as it does to believe that he exists.

5. "Tell me, what benefit is in forcing people (for example, taking children to church at an age where they aren't ready to decide) to attend church service which will force them to worship a god they have not seen?"

May I answer this question using this illustration: There are some people who are not happy with their lives and have expressed anger at why they exist. Would it be justifiable then to assert that sexual reproduction is wrong because it precludes the consent of the child to get born? Too ridiculous a question but one which should provoke the common sense in you and I to know that when a male and female decide to bring forth another human being on this planet; much as the new person will later have an independent mind to decide his lifestyle - whether to be a vegetarian or meat-eater, an atheist or theist - he will have no other way to start life apart from first being in the complete control and care of his parents. The *control* should gradually recede as the person grows older. This situation would not have been so if humans were born adults. But we have found ourselves in a world in which the very initiative and possibility of one's existence is in the control of his or her parents (i.e. they decided to make love and bring forth the child). Even after the child's birth, parents will still decide what food or type of schooling is best for the child. When this child grows into adulthood he may adopt an altogether different way of life (*become a vegetarian?*) and philosophy (*adopt theism or atheism?*) quite different from his parents. This is acceptable; what should be avoided is indoctrinating the child. In the words of the famous atheist, Richard Dawkins, "*What a child should never be taught is that you are a Catholic or Muslim child, therefore that is what you believe.*" For once, I agree with him. However, an over-stretch of this statement to mean "carrying your two year old child to church is abuse and indoctrination" would be an exaggeration which would lead to various absurdities of trying to get a child's consent on different issues which require personal choice (like the aforementioned case of birth-consent).

Wise and prudent parents do not force life's journey on their child; they will give him an opportunity to see the world as they see it, and later when he is mature enough to use discretion, give him space to be aware of other worldviews and let him decide. Is it not for this reason that Richard Dawkins is a man who was raised up by religious parents but himself later became an atheist, and Francis Collins, a son of "free-thinkers", was an unbeliever but later became a believer? Similarly, John Lennox had parents who, despite being Christians, encouraged him to think and analyse other world views.

In saying all this, however, I do not dispute your observation that some (or perhaps many) people force religion upon their children. Some religions are quite terrible on this. Some Christians are just as guilty when they tell people to believe because they just have to believe. In many churches today, sermons are so much filled with ignorance, even speaking against simple provable facts of science. But to be fair, try to recognise the fact that this is not representative of all Christians; there are still spiritual people (few as they may be) whose faith is not a blind one but one based on evidence and rigorous thought.

6. "The Bible is a white-man's book. Do you know what most whites have done after bringing religion to Africa? They have abandoned it, because it has done its job of being a mass enslavement tool."

This is quite an uninformed assumption; it arises from two things – misrepresentation (or ignorance) of history and a reaction to the terrible attitude of Europeans who mistreated blacks and abusing them as slaves. All this happened at a time when Christianity was every white man's garb. Now, however evil and hypocritical the settlers and colonialists may have been, to suppose that the Bible they upheld was merely used as a trick to enslave blacks is simply false. These whites did not invent religion in Africa. They were practising Christianity way before they started coming into Africa. Their misdeed, however, was to look upon Africans as inferior humans who were supposed to be evangelised but be kept under the subjection of their colonial rule. Be aware that there were a few missionaries who were enlightened and fought against this evil.

Another fallacy is to suppose that the Bible was produced by Europeans. This is quite an ignorant assumption but which unfortunately many uninformed people have accepted as fact. Europeans adopted the Bible just as Africans. The book is Jewish and hails from the Middle East. And is it not interesting to know that the

Jews (owners of the book) are actually the ones who have suffered terrible slavery and atrocities by gentile nations since time immemorial until the year 1948 when Israel became a nation? For 400 years they were in slavery under the ancient Egyptian empire. Then came Assyria, Babylon, Persia, Greece and the cruellest, Pagan Rome. In the nineteenth century, Hitler and Eichmann had his day of trying to exterminate the Jews. To-date there is still widespread anti-Semitism.

Now concerning the assertion that whites have abandoned religion after having used it as a mass enslavement tool; this is not true. But even if it were true, would it make a good argument against faith? Consider this: the oratorical Adolph Hitler became the leader of Germany through a democratic vote but later abused the system of governance turning it into a dictatorship regime. Now, does this mean democracy is bad because one evil man manipulated it for his ends? Not so. Beware that a wicked man will do everything at his disposal to further his interests – whether it is a Bible, Quran or even atheism.

7. *"If God exists why are there different religions? Christianity alone has thousands of denominations. Truth can't have different versions!"*

It is important to appreciate the fact that as long as we are dealing with things with a characteristic of subjectivity, different opinions will arise and effort should be made to analyse the most reasonable facts. Dismissing something on the basis of people's varying interpretations or perceptions of it is simply wrong. Two ants which happen to find themselves on *Manland* may have different perceptions of what the region is all about; the researcher in *Hairland* will have a thesis that will conclude that *Manland* is a huge thick dark forest which occasionally has *riceosaurus* monsters. The one in *Kneeland* may argue that *Manland* is an unstable curvature which continuously experiences concussions, and only becomes slightly stable at night. These different perceptions could be incomplete or wrong descriptions of man, but do not necessarily falsify the existence of the thing being described. Such discrepancies tend to become more problematic when dealing with non-tangible issues of love, trust, thoughts, faith, etc. Whilst a scientific instrument can measure the activity of adrenaline or blood pressure, it cannot measure or explain the thought of love that caused the surge. The blood or arteries are simply the medium through which the non-tangible occurs. The non-tangibles belong to a deeper dimension which cannot be fully captured or grasped by observation. As human beings we are trapped in the prison house of the body. In this state we only observe the world through two small windows called eyes and other senses. Thus, while our interpretation of physical phenomenon may be accurate and objective, non-tangibles have always posed an inherent difficulty and are hence usually subject to people's different perceptions. Even so, on the spectrum of the varied perceptions and interpretations, there are clearly sensible and justifiable insights just as there are downright ludicrous ones. Religion is not the only one faced with this problem. Scientists are also just as divided on quite a number of issues such as *bio-ethics*. However, their different views about these things does not equate to *"science being a lie because Truth has no different versions*". The 'different versions' are not in the universe or nature that science investigates but in the interpretations of the things being investigated. The universe will always be as it is, but our limited minds will always have different opinions about it. Differences in perception will always be there as long as you are dealing with the immeasurable, improvable or subjective. And it can't be denied that there are many things in this world which can't be subjected to empirical analysis and that is why faculties of philosophy have their place.

8. "Religious institutions do not pay tax, yet they are able to engage in political activities, public policies and everything to do with government. The money that religion alone has collected has the power to help a lot of people."

I have used the word 'religion' to generally refer to a profession of faith or simply 'belief in God'. My aim has been to give general arguments for the existence of God (*theism*) without going into particular doctrinal issues of what religion or which holy book is the true one. However, may I state that I speak as a Christian theist, and I answer questions in that capacity. This is not merely for the sake of it but because I find the Christian thought about God to be more coherent and truthful. Even so, when I call myself Christian and another person also calls himself Christian, it does not mean that we have the same behaviour or view about everything. Again, this in turn does not mean the Truth we are trying to understand is in 'different versions'.

I do not believe in the involvement of religion in politics. In my opinion that was the greatest error of the organised Christian Church back in the time of the Roman Empire when it embraced politics. I take Christianity as a way of life and not an organisation. Britannica Encyclopaedia has given a good historical exposition of how early Christians were not political or an organised entity as they are today. They were just people living a certain type of life. But gradually worldliness and politics crept into the Church and it turned out to be the worst tool of terror history ever knew.

Today much of what are called churches are nothing more than business and motivational centres. This compromise, in all the church meetings and conferences I have been invited to speak, locally and overseas, I have condemned in strongest terms.

9. **"Religion has been practiced for thousands of years and clearly it has not yielded any good results because it has failed to achieve peace. The only thing it has managed to do is robbing people of their little money by selling them delusions of heaven and hell..."**

Let me respond to this in two steps, (a) and (b) below:

(a) *"Religion has not yielded any good results"*: It is important to base opinions on facts and not mere perception. According to a study by the Barna Group:

The typical no-faith American donated just $200 in 2006, which is more than seven times less than the amount contributed by the prototypical active-faith adult ($1500). Even when *church-based giving is subtracted from the equation, active-faith adults donated twice as many dollars last year as did atheists and agnostics. In fact, while just 7% of active-faith adults failed to contribute any personal funds in 2006, that compares with 22% among the no-faith adults* (Barna Group, 2007).

And another report states:

The differences in charity between secular and religious people are dramatic. Religious people are 25 percentage points more likely than secularists to donate money (91 percent to 66 percent) and 23 points more likely to volunteer time (67 percent to 44 percent). And, consistent with the findings of other writers, these data show that practicing a religion is more important than the actual religion itself in predicting charitable behaviour. For example, among those who attend worship services regularly, 92 percent of Protestants give charitably, compared with 91 percent of Catholics, 91 percent of Jews, and 89 percent from other religions. (Brooks, 2003).

Another research, a comprehensive survey on religion in America, had this to say:

The differences between religious and secular Americans can be dramatic. Forty percent of worship-attending Americans volunteer regularly to help the poor and elderly, compared with 15% of Americans who never attend services. Frequent-attenders are also more likely than the never-attenders to volunteer for school and youth programs (36% vs. 15%), a neighbourhood or civic group (26% vs. 13%), and for health care (21% vs. 13%). The same is true for philanthropic giving; religious Americans give more money to secular causes than do secular Americans (Campbell and Putnam, 2010).

What do you make of these statistics?

(b) *"[Religion] has failed at achieving peace except of course robbing people of their little money by selling them delusions of heaven and hell."*

Kindly be aware that crooks, hypocrites and thieves exist in all sorts of *garb* – Christians, politicians, police officers, priests, pastors, accountants, etc. Would I be wrong to add atheists to this list?

In all this, one will be very mistaken to reckon that because there is widespread hypocrisy then sincere people do not exist. Just as there are some sincere politicians (few as they may be), sincere police officers (few as they may be), honest atheists (few as they may be), there are also sincere and honest people of faith, few as they may be! To abolish religion because of some cases of hypocritical pastors would seem similar to suggesting that we should abolish the police service because of some corrupt officials. Despite the many corrupt officials that could be there, the solution would be to eliminate the cancer of corruption in order to leave the service operating professionally. Do not doctors try all the best they can to focus efforts on eliminating the cancer which has eaten its way into the life of a person? Wouldn't the practice of physicians have been doomed if there was no knowledge of distinguishing the good life from the germ of death? Think about this: the atheist regime in the Soviet Union burnt and destroyed many churches, and Pol Pot of Cambodia killed his hundreds; so, would I be right to call you (an atheist) a violent person simply because you subscribe to atheism? Not so. There are many gentle and respectable atheists in this world just as there are many believers. Even if they were to be few, the fact still remains that there are some sincere believers, some who have even proclaimed and condemned the so-called 'Prosperity gospel' which is making business out of gullible people. Be reminded that to see *Jim* putting on a red coat, and then again see *Peter* putting on a red coat does not make Jim the same as Peter. Likewise a man in pink who stole from a mall does not make everyone who wears pink a thief. "The *sky* looks *grey*" and "That *cat* is *grey*" – is the sky the same as a cat? Ridiculous isn't it? So, to say "*Some or many religious people have 'robbed' money from people*" is true. But, "*all religious people 'rob' poor people*" is clearly misplaced and exaggerated.

NOTE: the following question was a response to the above explanation (from the same person who had asked the preceding question):

10. "…even if you say that religion has brought in good things. What things? Money?? Money is not what people need, people need mental freedom to exist."

Interesting statement. Money is simply a code (in form of paper or coins) which society uses to represent the power to access food, shelter and clothing. Take away these basic necessities from a person and see if he will have "*mental freedom*" to exist. The quotations I made about amounts of money religious people or charities give compared to atheists should be seen in the correct context. At the time of my typing this, there is a call from some villages in Malawi and Mozambique for financial assistance, following the devastating floods. I have been informed of one person who has died after contracting malaria. There is obviously mental depression among the affected families. The money I am sending, in itself, is not what will *"give them mental freedom to exist"*. However, that mental freedom is *encoded* in that money in that they will use it to buy food, clothing and shelter and that way another death can be avoided and hence some life will continue to *"exist"*. This reminds me of Siddhartha (the Buddha) who at one time realised that too much of fasting was affecting his mind in losing strength to reason; he therefore changed his approach by avoiding extremes in life.

11. *"People become Christians because they fear hell fire and want to escape it by going to heaven. All these are delusions."*

Any person who truly believes and loves God knows that heaven is not a form of *"Fire Insurance"*. While it is true that some Christians try to evangelise people by scaring them with hell, not everyone subscribes to such old medieval tricks.

Concerning heaven and hell being delusions, the debate can't start from this doctrinal point. The logical start point of the discussion should be with the question of whether God or the supernatural exists. If the answer is 'yes', then the existence of hell and heaven becomes possible.

12. "One thing you must also know is that, it is not possible for a supernatural to come in to a natural world."

Time began when the smallest fraction of matter came into existence. This is called the beginning and it is unavoidable as the *Borde-Guth-Vilenkin* theorem and other models have convincingly demonstrated. And much as we can try to measure the tiniest microseconds of events, the greatest difficulty is in what lies beyond the existence of matter. In that 'space' – let me call it '*the beyond*' - all known rules of reality break down. Hawking tried to work out a model which uses imaginary numbers but which can't convert back to real ones; this is simply because in *the beyond* everything including time and space ceases. That realm is beyond *natural*; yes, it is *super-natural*. And it is from there that something came into existence. By implication then the natural exists within the supernatural. Thus, the supernatural is the underlying force by which everything lives and exists. The recent largest scientific study ever conducted on life after death experiences (see **Appendix II**) has given convincing evidence that when a body is dead, life and human consciousness continues to exist. What else can this mean but that there is something more to this world than the natural?

13. "Why do we need God when evolution has explained how things became what they are?"

The word evolution is now so commonplace in vocabulary that its meaning to an uninformed person has become vague. For instance one can say something like "the evolution of democracy" or "the city has evolved". In such statements the word evolution is simply used to generally refer to change that has occurred in an environment, place or thing. Of course this context of usage would make everyone an evolutionist; we all perceive changes in surroundings, demographics or stages of growth from an embryo to a full grown person. Likewise there are variations which can exist in a particular animal kind; dogs for example come in different types with different features. Surely some sort of evolution occurs, for example, due to *mutation* or *natural selection*. However, confusion begins when someone posits that one *kind* of a specie originated from another kind, and through this, tracing human beings' ancestry to monkeys and further down to fish and onto to a microbe. Sir Fred Hoyle, the British astrophysicist and mathematician, could not be more correct when he said, *"Common sense would suggest, the Darwinian Theory is correct in the small, but not in the large. Rabbits come from other slightly different rabbits, not from either [primeval] soup or potatoes"* (Hoyle, 1987, p.7).

Surely evolution *within* species occurs; for example different rabbits exist, some with longer ears and others shorter; it is undisputable that such changes can occur as a result of different kinds of rabbit mating or due to adaption to environments. Such small changes that occur within species are well known as *microevolution*. *Macroevolution* on the other hand refers to the *assumed* major changes in species from one kind to another – from primeval soup to a microbe on to an ape and on to a human being. Macroevolution rests on the premise that micro changes in organisms will over a large span of time ultimately lead to major changes. When this theory was first presented by Darwin it sounded too absurd to people, but with repeated emphasis and elaboration over time, it gained momentum until some adherents have become so convinced of it as fact: "*evolution is fact, not theory …It is a fact that all living forms come from previous living forms. Therefore, all present forms of life arose from non-birds and humans from non-humans*" (Lewontin, 1981, p.559). It is further said by many proponents that now because we have an explanation of what led to the rise of differences in species – evolution through a gradual process of natural selection – the Genesis story of a divine creator has no longer any place. But the flaw in the logic of comparing the alternative of a creator and of evolution cannot go unnoticed.

Faulty logic

Huxley (1960) stated that "*in the evolutionary scheme of thought there is no longer either need or room for the supernatural. The earth was not created, it evolved. So did all the animals and plants that inhabit it, including our human selves, mind and soul as well as brain and body*" (cited in Lennox, 2009, p.87). These words make evolution compete against God (and please note the overreach of evolution, in Huxley's words, to now account for the beginning of the earth and not just living things); Huxley makes evolution (a *mechanism* of how diversities in living things arose) contend against belief in God (*agent* or maker of the living things) as the creator. Logical error detected:

Consider this: which is greater, the *number* '2' or the *letter* 'h'? The problem here is not with what answer to give but with the question itself. The question is comparing items belonging to different categories – the numerical scheme category and the alphabetical category; it is a case of *category mistake*. Comparing God and evolution suffers the same mistake. Evolution and God are not competing explanations: the former is a *mechanism* that explains how diversity in living organisms arose through natural selection and the latter an *agent* who brought about the life. Evolution does not explain what caused the beginning of life, but merely attempts to explain how the diversity of its forms arose[26]. Thus, evolution can only be compared to God if it has an agent (other than God) which caused life to begin to exist. An illustration may save a thousand words:

A cook is asked a question:

> ***How*** *did this nice cake come to be* (this question asks for mechanism)*?*
>
> Answer: *The recipe involved preparing the dough, putting it in an oven, and later icing it with cream.*

Second question:

> ***Who*** *made this cake?* (this question asks for the agent by whom the cake came to be; the rationale is that the process of dough preparation, heating and icing cannot happen on its own, i.e., without involvement of an intelligent agent).

Answer: *Jack did it.*

---

[26] It is no wonder that some scientists like Francis Collins believe both in God and evolution: in their worldview God is the *agent* that caused the *mechanism* of evolution. This would not be expected if *evolution* and *creator God* were mutually exclusive explanations.

Third question:

> *Of the two answers below, which one gives a more sensible explanation?*
>
> a) *Making the cake involved preparing the dough, putting it in an oven, and later icing it with cream.*
> OR,
>
> b) *Jack prepared the cake.*

There is clearly no problem with the first two questions but one in the third: no comparison can be drawn between alternatives (a) and (b) of the Third Question. Q3(a) asks about the mechanism (*recipe*) used to prepare the cake and (b) about the agent (*Jack*) who prepared the cake. The *recipe* and *Jack* cannot logically compete as explanations. The question will only make sense if there is an alternative person to compare with (say, Peter whom everyone knows has never baked before) or an alternative recipe (say, one which may or may not have led to producing a cake in question). The third question can thus be re-written as follows:

> *Of the two answers below, which one gives a more sensible explanation as to how the cake was made?*
>
> a) *The cake was prepared by first making the dough, heating it in the oven and later icing it,*
> *Or*
>
> b) *The cake was prepared by first making the dough, cooling it in the freezer and later icing it?*

It can now be clearly seen that both alternatives involve mechanism, and hence can be logically compared. So, the second part of the third question can also be re-written as follows:

Of the three answers below, which one gives a more sensible explanation as to who or what made the cake?

a) Jack prepared the cake?

b) The cake made itself

c) The cake has been existing eternally, without beginning or end.

Options (b) and (c) are clearly nonsensical and extending the stature of Jack or increasing the size of the cake until it feels the universe can never rationalize the absurdities!

Recommended books on the subject of the interface between science, faith and morality:

i) God's Undertaker: *Has science buried God?*
    By John Lennox
    (Visit: www.johnlennox.org)
    ISBN: 9780825479120

ii) On Guard: *Defending your faith with reason and precision*
    By William Lane Craig
    (Visit: www.reasonablefaith.org)

iii) The Dawkins Delusion: *Atheist fundamentalism and the denial of the divine*
    By Alsiter McGrath and Joanna Collicutt McGrath
    ISBN: 9780281059270

iv) Dawkins God: *Genes, memes, and the meaning of Life*
    By Alsiter McGrath
    ISBN: 9781405125383

v) Not By Chance: *Shattering the modern theory of evolution*
    By Lee Spetner
    ISBN: 9781880582244

vi) The Language of God: *A scientist presents evidence for belief*
    By Francis Collins
    ISBN: 9781847390929

vii) Is God a Moral Monster? *Making sense of the Old Testament God*
    By Paul Copan
    ISBN: 9780801072758

viii) God, Freedom and Evil
    By Alvin Plantinga
    ISBN: 0802817319

Appendices

I. Nuclear efficiency

One kind of an atom can exist in different forms. Its different versions are called *isotopes*. For example Hydrogen exists in the following forms:

- *Protium*, written as ^1H because it consists of 1 proton and has no neutrons.
- *Deuterium* written as ^2H because it consists of 1 proton and 1 neutron which makes a mass of 2.
- *Tritium*, ^3H, 1 proton, 2 neutrons hence making up the mass of 3.

As you may have noticed above, the number of protons plus the number of neutrons of an atom make up its mass. Nuclei of two or more atoms can collide at high speeds and join (fuse) to form new atomic nuclei and in the process release energy. The process is called *nuclear fusion* and it is responsible for the generation of heat energy of the sun. Nuclear fusion occurs when ^2H fuses with ^3H to form ^4He (Helium) - See illustration below:

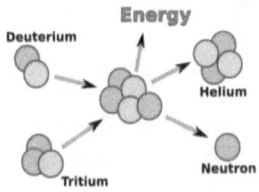

Hydrogen 2 fusing with Hydrogen 3

Note that the fusion of ^2H and ^3H haven't resulted into *mass 5* but *mass 4*. The other unit of mass has been released as heat energy. This released energy is 0.7% (or 0.007) of the mass that was supposed to be the total of the protons and neutrons that formed the helium nucleus. Thus the fuel that powers the sun – Hydrogen - converts 0.007 of its mass into energy when it fuses into helium. This number – 0.007 - is crucial and is denoted by the symbol $\mathcal{E}$.

Experts have explained stunning facts concerning $\mathcal{E}$. Here is what Sir Martin Rees, a British astrophysicist has explained concerning how a slight adjustment to the value of $\mathcal{E}$ would have inhibited life.

What would have happened if $\mathcal{E}$ were weaker?

> *If the nuclear 'glue' were weaker, so that $\mathcal{E}$ were 0.006 rather than 0.007, a proton could not be bonded to a neutron and deuterium would not be stable. Then the path to helium formation would be closed off* (Rees, 2000, p.49).

And if $\mathcal{E}$ were stronger?

> *We couldn't have existed if $\mathcal{E}$ had been more than 0.008, because no hydrogen would have survived from the Big Bang...if $\mathcal{E}$ were to have been 0.008, then two protons would have been able to bind directly together. This would have readily happened in the early universe, so that no hydrogen would remain to provide the fuel in ordinary stars, and water could never have existed.* (p.49).

II. First hint of 'life after death' in biggest ever scientific study.

By Sarah Knapton, 7 October 2014
(news article reproduced from *The Telegraph*).

Death is a depressingly inevitable consequence of life, but now scientists believe they may have found some light at the end of the tunnel. The largest ever medical study into near-death and out-of-body experiences has discovered that some awareness may continue even after the brain has shut down completely. It is a controversial subject which has, until recently, been treated with widespread scepticism.

But scientists at the University of Southampton have spent four years examining more than 2,000 people who suffered cardiac arrests at 15 hospitals in the UK, US and Austria. And they found that nearly 40 per cent of people who survived described some kind of 'awareness' during the time when they were clinically dead before their hearts were restarted. One man even recalled leaving his body entirely and watching his resuscitation from the corner of the room. Despite being unconscious and 'dead' for three minutes, the 57-year-old social worker from Southampton, recounted the actions of the nursing staff in detail and described the sound of the machines.

"We know the brain can't function when the heart has stopped beating," said Dr Sam Parnia, a former research fellow at Southampton University, now at the State University of New York, who led the study. *"But in this case, conscious*

awareness appears to have continued for up to three minutes into the period when the heart wasn't beating, even though the brain typically shuts down within 20-30 seconds after the heart has stopped. The man described everything that had happened in the room, but importantly, he heard two bleeps from a machine that makes a noise at three minute intervals. So we could time how long the experienced lasted for. He seemed very credible and everything that he said had happened to him had actually happened."

Of 2060 cardiac arrest patients studied, 330 survived and of 140 surveyed, 39 per cent said they had experienced some kind of awareness while being resuscitated. Although many could not recall specific details, some themes emerged. One in five said they had felt an unusual sense of peacefulness while nearly one third said time had slowed down or speeded up. Some recalled seeing a bright light; a golden flash or the Sun shining. Others recounted feelings of fear or drowning or being dragged through deep water. 13 per cent said they had felt separated from their bodies and the same number said their senses had been heightened.

Dr Parnia believes many more people may have experiences when they are close to death but drugs or sedatives used in the process of resuscitation may stop them remembering. *"Estimates have suggested that millions of people have had vivid experiences in relation to death but the scientific evidence has been ambiguous at best. Many people have assumed that these were hallucinations or illusions but they do seem to corresponded to actual events. And a higher proportion of people may have vivid death experiences, but do not recall them due to the effects of brain injury or sedative drugs on memory circuits. These experiences warrant further investigation."*

Dr David Wilde, a research psychologist and

Nottingham Trent University, is currently compiling data on out-of-body experiences in an attempt to discover a pattern which links each episode. He hopes the latest research will encourage new studies into the controversial topic.

"Most studies look retrospectively, 10 or 20 years ago, but the researchers went out looking for examples and used a really large sample size, so this gives the work a lot of validity. There is some very good evidence here that these experiences are actually happening after people have medically died. We just don't know what is going on. We are still very much in the dark about what happens when you die and hopefully this study will help shine a scientific lens onto that."

The study was published in the journal Resuscitation. Dr Jerry Nolan, Editor-in-Chief at Resuscitation said: *"Dr Parnia and his colleagues are to be congratulated on the completion of a fascinating study that will open the door to more extensive research into what happens when we die."*

III. Halo photographed in Houston Texas.

In 1950 a halo of light appeared over the head of Rev. Branham. He had previously claimed to have always seen the light following him from the time he was a young boy. Many other people, some still living today, testified of having seen the supernatural light. At one time a newspaper reported of an incident at Ohio River of a baptismal service attended by hundreds of people where a number of them screamed and others fainted at the appearance of a glow of light atop the evangelist. Obviously this can easily be dismissed by sceptics by concluding that it should have been hysteria, which is typical of modern charismatic meetings where much of what is called supernatural is just faked. However, when a scientific instrument of a camera one day captured the halo, one sceptic and a critic remarked that a camera doesn't know psychology or tricks of hysteria. George J. Lacy, examiner of questioned documents, an FBI agent examined the image and was astonished to confirm the paradox. This was his report:

George J. Lacy
Examiner Of Questioned Documents
Shell Building
Houston, Texas, U.S.A.

REPORT AND OPINION

Re: Questioned Negative January 29, 1950

On January 28, 1950 at the request of Rev. Gordon Lindsay, who was representing Rev. William Branham of Jeffersonville, Indiana, I received from the Douglas Studios of 1610 Rusk Avenue in this City, a 4 x 5 inch exposed and developed photographic film. This film was purported to have been made by the Douglas Studios of Rev. William Branham at the Sam Houston Coliseum in this city, during his visit here the latter part of January, 1950.

Request

Rev. Lindsay requested that I make a scientific examination of the aforesaid negative. He requested that I determine, if possible, whether or not in my opinion the negative had been "retouched" or "doctored" in any way, subsequent to the developing of the film, that would cause a streak of light to appear in the position of a halo above the head of Rev. Branham.

Examination

A macroscopic and microscopic examination and study was made of the entire surface of both sides of the film. Both sides of the film were examined under filtered ultra-violet light and infra-red photographs were made of the film.

The microscopic examination failed to reveal retouching of the film at any place whatsoever by any of the processes used in commercial retouching. Also, the microscopic examination failed to reveal any disturbance of the emulsion in or around the light streak in question.

The ultra-violet light examination failed to reveal any foreign matter, or the result of any chemical reaction on either side of the negative, which might have caused the light streak, subsequent to the processing of the negative.

The infra-red photograph also failed to disclose anything that would indicate that any retouching had been done to the film.
The examination also failed to reveal anything that would indicate that the negative in question was a composite negative or a double exposed negative.

There was nothing found which would indicate that the light streak in question had been made during the process of development. Neither was there anything found which would indicate that it was not developed in a regular and recognized procedure. There was nothing found in the comparative densities of the highlights that was not in harmony.

Opinion

Based upon the above described examination and study I am of the definite opinion that the negative submitted for examination, was not retouched nor was it a composite or double exposed negative.

Further, I am of the definite opinion that the light streak appearing above the head in a halo position was caused by the light striking the negative.

Respectfully submitted,

George J. Lacy.

See details about the story surrounding this report and the ministry and life of William Branham in Jorgensen (2008).

A thought to consider, If this recent scientifically validated miracle is doubted, is there chance for a sceptic to believe in the story of a carpenter's son who arose from death two thousand years ago?

IV. Dating of Fossils

Various methods are used to date historical artifacts. Some methods determine the age of fossil in relative terms and others in absolute.

In relative terms a fossil embedded in a sediment, for example, is dated in comparison to different strata of rock, taking lower layers to be older than the upper ones. In other cases layers are compared to others in different locations. With this method, fossil content that matches with similar ones in another strata of a different place are said to belong to the same age.

Absolute dating is more specific and involves measuring the regular decay of radioactive (parent) *isotopes* into other *daughter* isotopes. The ratio between parent and daughter isotopes is used to approximate the age of an artifact. Various radiometric dating methods exist; Carbon-14 dating is the common one but which can only measure organic material less than 50,000 years old. Older artifacts require isotopes with longer *half-lives* such as Potassium-40 or Rubidium-87.

Various experiments have proved these methods to be reliable. Unfortunately many proponents of a 6000 year old earth seem to be either unaware of these scientific dating methods or give undue criticism. Wiens(2002) correctly notes that *"many Christians have led to distrust radiometric dating and are completely unaware of the great number of laboratory measurements that have shown these methods to be consistent"* and that *"many people have been led to be skeptical of dating without knowing much about it."* See Dalrymple (1991), Young (1988) and Wiens (2002) for details on this subject.

A.C. Phiri is pastor of *Believers' Assembly*, a non-denominational church in Zambia, where Christians seeking in-depth Bible study, prayer and fellowship gather to worship. Andrew has spoken and taught in various meetings and conferences across Africa and overseas including Singapore, the Philippines, India, Cambodia, Namibia, Uganda, Zimbabwe, Malawi and Mozambique. He takes keen interest in missionary work and helping the poor in rural villages of Africa. In noticing the lack of knowledge about the interface between faith and science, he has for the past two years concentrated on teaching this subject.

References

Alvarez L.W. , Alvarez, W., Asaro, F., and Michel, H.V. (1980). *Extraterrestrial Cause for the Cretaceous-Tertiary Extinction* [Online]. Science, New Series, Vol. 208, No. 4448. pp. 1095-1108. Available from: http://www.jstor.org/stable/1683699 [Accessed 11th June 2013].

Bakos, N. (2013). Humility now! The miseducation of Jackson Diehl. *Foreign Policy*. [Online]. Accessed from: http://foreignpolicy.com/2013/04/02/humility-now/ [Accessed: 15th June, 2015].

Barna Group (2007). *Atheists and Agnostics Take Aim at Christians*. [Online] Available from: https://www.barna.org/barna-update/faith-spirituality/102-atheists-and-agnostics-take-aim-at-christians#.VWl_ds-qqko [Accessed: 30th May, 2015].

Britannica Encyclopaedia (2015). *Soviet invasion of Afghanistan*. [Online] Available from: http://www.britannica.com/event/Soviet-invasion-of-Afghanistan [Accessed: 15th June, 2015].

Brooks, A.C. (2003). *Faith and Charitable Giving*. [Online]. Available from: http://www.hoover.org/research/religious-faith-and-charitable-giving. [Accessed: 30th May, 2015].

Campbell , D. and Putnam, R. (2010). *Religious People are Better Neighbors*.[Online]. Available from: http://usatoday30.usatoday.com/news/opinion/forum/2010-11-15-column15_ST_N.htm [Accessed: 30th May, 2015].

Collins, F. (2007). *The Language of God*. London: Simon and

Schuster.

Copan, P. (2011). *Is God a Moral Monster; Making Sense of the Old Testament.* USA: Baker.

Craig, W.L. (2010). *The New Atheism and Five Arguments for God.* [Online] Available from: http://www.reasonablefaith.org/the-new-atheism-and-five-arguments-for-god [Accessed 30th March 2015].

Crick, F. (1994). *The Astonishing Hypothesis – The Scientific Search For The Soul.* London: Simon and Schuster.

Dalrymple, B.G.(1991). *The Age of the Earth.* Stanford, CA: Stanford University Press.

Davies, P. (2006). *The Goldilocks Enigma.* London: Penguin.

Davies, P. (2006). *The Goldilocks Enigma.* London: Penguin.

Dawkins, R. (1996). *River out of Eden; A Darwinian view of life.* New York: Basic Books.

Dawkins, R. (2006). *The God Delusion.* Boston: Houghton Mifflin.

Gitt, W. (2000). *In the Beginning was Information.* Germany: CLV.

Goswami, A. (2012). *God Is Not Dead: What Quantum Physics Tells Us about Our Origins and How We Should Live.* Mumbai: Jaico Publishing House.

Harrison, E. (1985). *Masks of the Universe.* New York. Macmillan.

Hawking, S.(1998). *Brief History of Time.* New York: Bantam Press.

History Channel Website (2013).*Why Did The Dinosaurs Die Out?* [Online]. Available from: http://www.history.com/topics/why-did-the-dinosaurs-die-out [Accessed 9th June 2013].

Hoyle, F. (1981). The Universe: Past and Present

Reflections. *Engineering and Science.* November.

Huxley, J. (1960). *Evolution after Darwin,* Sol Tax ed. Chicago: University of Chicago Press.

Jorgensen, O. (2008). *Supernatural: The Life of William Branham*: (Book Three: The man and his Commission). Arizona: Tucson Tabernacle.

Kline, M. (1980). *Mathematics: the loss of certainty.* New York: Oxford University Press.

Knapton, S. (2014). *First Hint of 'Life After Death' In Biggest Ever Scientific Study.* [Online] Available from: http://www.telegraph.co.uk/news/science/science-news/11144442/First-hint-of-life-after-death-in-biggest-ever-scientific-study.html [Accessed: 20th April, 2015].

Leaky, R. (1990). *African Fossil Man.* In: Ki-Zerbo, J. (ed).*UNESCO General History of Africa.* Methodology of African History.Vol.1: UNZA Press.

Lennox, J. (2009). *God's Undertake; Has Science Buried God?* Oxford: Lion Hudson.

Lennox, J. (2011). *God and Stephen Hawking.* Oxford: Lion Hudson.

Lewis, C.S. (1947). *Miracles.* New York: HarperCollins.

Lewis, C.S.(1942). *Mere Christianity.* New York: HarperCollins.

Lewontin, R.C.(1981). Evolution/Creation Debate: A time for Truth. *Journal of Bioscience.*31(8).p.559.

Manson, N.A. (2003). 'Introduction' in *God and Design: the teleological argument and Modern science.* (Ed). London: Routledge.

Medawar, P. (1979). *Advice to a Young Scientist.* London: Harper and Row.

Pasachoff, J.M. (2009). *Steady-State Theory.* Redmond WA:

Microsoft Corporation.

Plantinga, A. (1974). *God, Freedom, and Evil.* USA: Eerdmans.

Polkinghorne, J. (1986). *One World.* London: SPCK.

Rees, M. (2000). *Just Six Numbers: The Deep Forces that Shape Our Universe.* New York: Basic Books.

Rees, M. (2003). *Out Cosmic Habitat.* London: Phoenix.

Smith, Q. (1993). The Wave function of a Godless Universe in *Theism, Atheism, and Big Bang Cosmology.* Oxford: Clarendon Press.

Stern, J. and Berger, J.M. (2015). *ISIS: the state of terror.* London: Williams Collins.

Swineburne, R. (1995). *Is there a God?* Oxford: Oxford University Press.

Tolle, E. (2005). *A New Earth: Awakening to your Life's Purpose.* London: Penguin.

Wiens, R.C.(2002). *Radiometric Dating: A Christian Perspective*: The American Scientific Affiliation.

Young, D.A.(1988). *Christianity and the Age of the Earth.* California: Artisan Sales.

Index

Afghanistan 23, 156
al Qaeda 23, 25
Alvarez, Luis 125, 126, 156
Anders, William..........113
anomie...........................82
anthropic principle......57
Aquinas, Thomas 54
Babylon.........................32
Bakos, Nada 25, 156
Barna Group.... 135, 136, 156
Bavaria 81
BGV Theorem..............46
Big Bang ...37, 40, 41, 46, 48, 50, 54, 57, 60, 106, 123, 128, 148, 159
Boko Haram.... 16, 21, 29
Borde-Guth-Vilenkin theorem...................140
Borman, Frank...........113
Branham, William 107, 108, 151, 152, 158
Broken Hill.................123
cake illustration of causation 143
carbon 62
Carbon 14....................154
Carbon Resonance......62
causal relation 22, 25
civilisation and human morality 32
consciousness.. 37, 69, 70

Copan, Paul.... 16, 30, 33, 157
Craig, William 15, 47, 51, 53, 86, 157
Crick, Francis...... 69, 157
Darwin, Charles.........117
Davies, Paul.... 63, 64, 65, 157
Dawkins, Richard.15, 27, 29, 30, 70, 93, 119, 130, 157
Dennet, Daniel ... 15, 119
Deoxyribonucleic Acid ....*See* DNA
Deuterium 147
DNA . 39, 42, 71, 85, 128
Douglas Studios.........152
electrons......61, 70, 93, 94
entropy....................51, 82
evolution...... 87, 143, 158
expansion of the universe 48, 57, 59
faith...... 15, 16, 37, 39, 40
faulty logic 142
fine-tuned expansion .. 59
finite space boundary.. 46
Fleming, Alexander.....22
Flew, Anthony.............42
Ford motor car illustration of agent and mechanism 42
Free the Body Culture 81
free will 69, 71

Galapagos islands 117
genocide 29
Gitt, Werner 157
Goswami, Amit 63, 77, 157
gravity .. 49, 52, 57, 58, 60
Hawking, Stephen 51, 52, 57, 60, 120, 140, 157, 158
Hearing the small still voice 111
helium 62
History Channel 126, 157
Hitler, Adolf .. 29, 85, 132
Holocaust 86
Homo erectus 123
Hoyle, Fred ... 49, 62, 141
Human Genome Project 41
immorality 88
iridium 125
ISIS 16, 29
isotopes 154
Jackson, Raymond 125
Jeremiah's prophecy giving data about demise of prehistoric era 124
Kalam Cosmological Argument 47
Kepler, Johannes 37
Lacy, George J. .. 152, 153
Leakey, Richard 123
Lennox, John 52
Lewontin 142
life after death scientific study 149, 158
Lovell, James 113
Macroevolution 141
Mailoni Brothers 30
materialism .. 69, 71, 84, 85
McGrath, Alister ... 15, 27
mechanism versus agency 142, 143, 144
microevolution 141
microwave background radiation 41
morality 32, 69, 87
mujahideen 24
Mulenga, Teddy 14, 15, 21
multiverse 64, 65
NASA 105, 127, 128
neutrons 61, 62
Noah 31
Northern Rhodesia ... 123
nuclear efficiency 60
nucleus 61, 62, 148
nudity 81, 82
Objective moral values 86
Old Testament violence 29
Oscillating model 50
Oxygen 63
Pachycephalosaurus .. 122
Polkinghorne, John 71
Polkinhorne, John . 15, 65
prehistory 123
primeval soup 141
Principia Mathematica 41
Proof, *comparing with*

evidence 35, 40
Protium.......................147
protons............. 58, 61, 62
quarks.............................61
Rees, Martin 65, 148, 159
replenish (Genesis 1:28)
123
resonance.......................62
SETI............................128
Shklovsky, Joseph 125
Soviet Union's invasion
 of Afghanistan.........23
Specified complexity...38
Steady-state model......49

strong nuclear force....61
Swinburne, Richard.....43
terrorism in Iraq..........25
Tolle, Eckert 28
Tritium........................147
University of
 Southampton..........149
weak force 58
Wiens, Roger C..........154
Yahweh......29, 30, 31, 32
Young, Davis 154
Yucatan Peninsula.....126
Zarqawi.........................25

Notes

Notes

Notes

Notes

www.ingramcontent.com/pod-product-compliance
Lightning Source LLC
Chambersburg PA
CBHW031055180526
45163CB00002BA/844